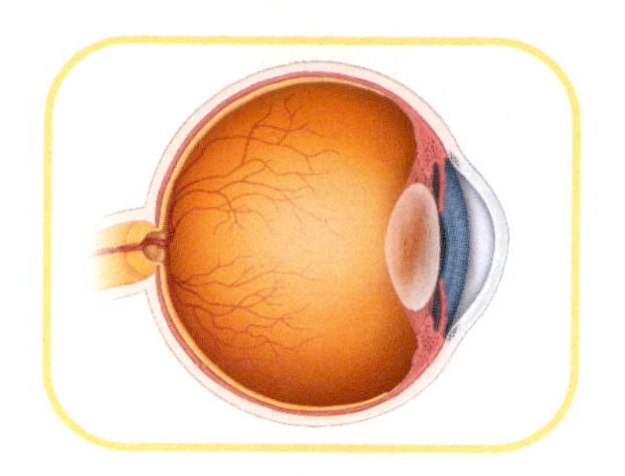
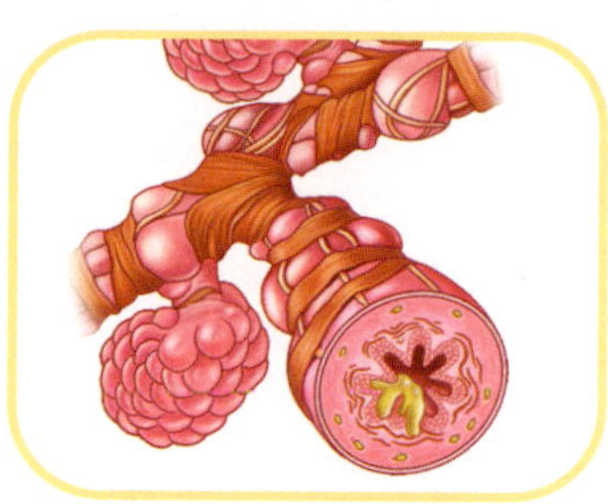
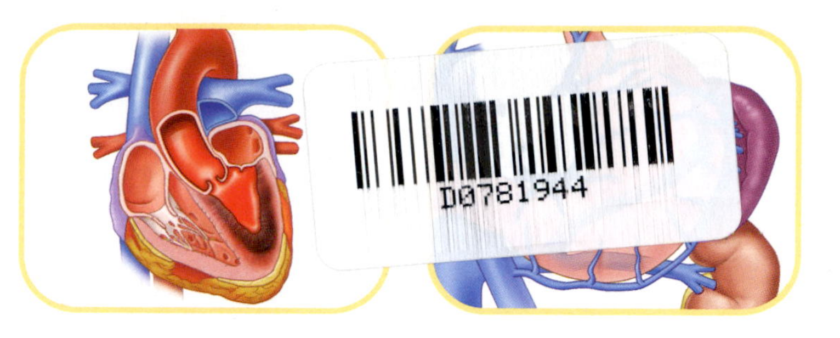

Case Mysteries in Pathophysiology

Patricia J. Neafsey, RD, PhD
University of Connecticut

Morton Publishing Company
925 W. Kenyon Avenue, Unit 12
Englewood, CO 80110

www.morton-pub.com

Book Team

Publisher	Douglas N. Morton
Biology Editor	David Ferguson
Editorial Assistant	Desireé Coscia
Cover & Design	Bob Schram, Bookends, Inc.
Illustration	Imagineering Media Services, Inc.
Copyeditor	Carolyn Acheson
Composition	Ash Street Typecrafters, Inc.

Printed in the United States of America

10 9 8 7 6 5 4 3 2 1

ISBN-10: 0-89582-769-7

ISBN-13: 978-089582-769-2

Library of Congress Control Number: 2009926258

Preface

Health science students are eager to use their knowledge of anatomy, physiology, and pathophysiology in clinical applications. The first clinical cases that students encounter are likely to remain indelible in their minds because of the novelty, the interpersonal aspects, and the excitement that come with problem-solving. But anatomy, physiology, and pathophysiology courses are offered during pre-clinical years or during early clinical rotations, when opportunities for students to care for a cardiac or renal patient may be limited. I designed *Case Mysteries in Pathophysiology* to help bridge that gap. This book offers opportunities to apply critical thinking skills to case studies while students are still in those pre-clinical courses.

Case Mysteries

Case Mysteries in Pathophysiology takes a problem-based learning approach. Each case presents a realistic clinical mystery—many with an interesting "twist" at the end. The case titles and associated icons are designed to help students remember details about the patient featured in each case. Normal findings (e.g., lab values, ECGs, X-rays, MRIs, CT scans, pathology slides) are presented next to the patient findings. Even though they are not expected to interpret ECGs or other images correctly at this stage of training, students will be able to identify abnormalities in these scans.

Students may work independently or with others to solve the case mysteries. They may find that working in groups is a stimulating challenge that fosters teamwork—an essential skill, as clinicians at all levels must work together to provide the best medical care possible. These case mysteries also will help students develop clinical skills in preparation for their rotations in coming years.

As all books evolve over time, I welcome your feedback on how this book can be improved. Contact me at patricia.neafsey@uconn.edu

DISCLAIMER

This book was designed strictly as a learning tool for health science students. Dosages, method and duration of use, and contraindications should be verified by checking with the appropriate drug manufacturers and prescribers. The publisher and author assume no liability for injuries or harm that may take place from the application of therapies detailed in this publication.

 NOTE TO INSTRUCTOR

There are two different versions of *Case Mysteries in Pathophysiology*. One version includes the answers to the student questions and the other version does not (answers to the student questions are available online). Please be sure to submit the correct ISBN to the bookstore to ensure your students receive the version you have selected.

Case Mysteries in Pathophysiology (including answers to student questions) **ISBN 978-0-89582-769-2**

Case Mysteries in Pathophysiology (without answers to student questions) **ISBN 978-0-89582-824-8**

Acknowledgments

I am especially appreciative of David Ferguson, Biology Editor at Morton Publishing Company for the opportunity to publish these case mysteries. His visionary approach to blending problem-based learning methods with superior quality visuals is a Morton hallmark. Thank you to Jay Zimmer of South Florida Community College, Jody Johnson of Arapahoe Community College, Dr. Carolyn White of the University of South Alabama, Alona Angosta of the University of Nevada-Las Vegas, Dr. Mary Knudtson of the University of California-Irvine, Carey Bosold of Arkansas Tech University, Dr. Jo A. Voss of South Dakota State University, Dr. Chris McConnell of South Florida Community College, and John Fishbeck of Ozarks Technical Community College, who provided many constructive comments (all of which I attempted to heed). Desireé Coscia, Publisher's Assistant, was delightful to work with on image acquisition, permissions, references, and editing. Many thanks to Carolyn Acheson for copyediting the manuscript. Joanne Saliger of Ash Street Typecrafters was efficient and accurate in her work with layout and image rendering. Many special thanks to the talented artists at Imagineering for their illustrations. Finally, I am grateful to the more than 3,000 students that have joined me in early morning classes to explore the mysteries of pathophysiology for the past 20 years.

Thank you to my family for their love, support, and patience.

Contents

1

Neck Hygiene

Charlene Woodstone is a 13-year-old girl in the 8th grade in a large junior high school in Texas. She is in the nurse's office in distress. "Ms. Jones told me to see you because I haven't got the dirt off my neck. I am *so* embarrassed! She talked to me in homeroom in front of everybody! I told her I *always* wash my neck, but she wouldn't even look at my neck up close or even *listen* to me!"

The information the school nurse has about Charlene is as follows:

Age: 13
Height: 5'3" (63 inches)
Weight: 171
Age at menarche: 10
Last period: "three months ago? periods not regular but frequent pelvic pain"
Blood pressure: 144/92
Pulse: 80

When asked if Charlene has any other concerns, she replies, "I dunno. I have hair where I shouldn't—like on the sides of my jaw and on my thumbs and toes. And my acne seems to be getting worse. I feel like a freak show." Charlene keeps her head down so her long hair covers the sides of her face. "The only thing I have going for me is singing soprano in the choir—but my choir teacher is on my case, too, because I sing with my head down."

The nurse notes that the "dirt" on Charlene's neck is actually an area of increased pigmentation (hyperpigmentation) that has a thickened, rough surface (hyperkeratosis) that looks like velvet.

The nurse says, "Charlene, I'm going to call your mother and ask that she take you to a special doctor. I think you can be helped so the appearance of your skin improves and the hair on your face goes away."

Below is a photo of Charlene's face and neck:

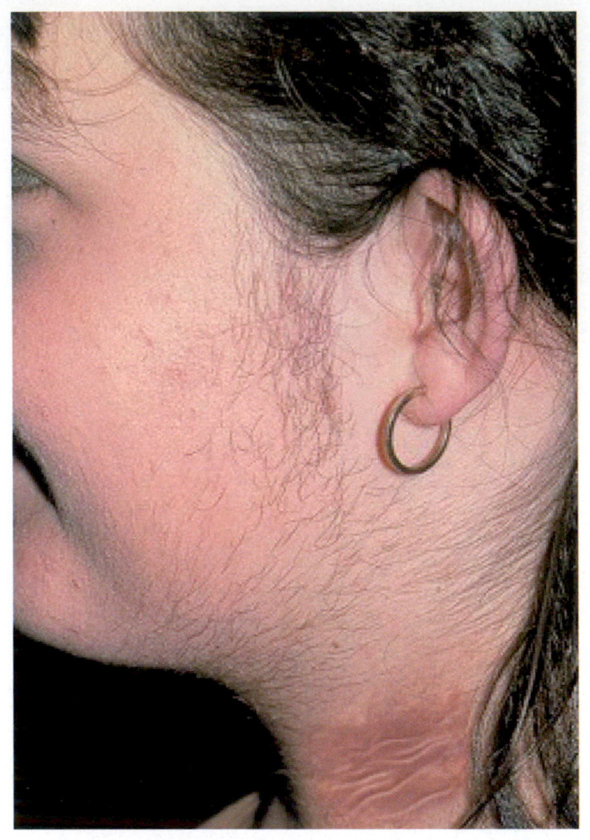

Original image was published in *The Journal of American Academy of Dermatology*, Vol. 55, Lindy P. Fox, *Diseases of the Skin: A Color Atlas and Text*, 2nd Edition, Copyright Elsevier 2006.

Questions

Name ______________________________ Section ________

1 Assess Charlene's height and weight by calculating her body mass index (BMI) using the online calculator at: http://www.nhlbisupport.com/bmi/. Comment on your findings.

2 Assess Charlene's blood pressure reading.

3 What problem is causing the hyperpigmentation and hyperkeratosis on Charlene's neck?

4 What tests would diagnose the underlying causes of this problem?

5 What problem is causing Charlene to have irregular periods and pelvic pain, and to grow hair in unusual locations?

Charlene's new doctor orders blood tests and an ultrasound. Below are selected results.

Fasting blood sample:

	Charlene's values	Normal values
Glucose	122 mg/dL	70–100 mg/dL
Triglycerides	275 mg/dL	15–150 mg/dL
LDL cholesterol	178 mg/dL	0.0–99.0 mg/dL
HDL cholesterol	35 mg/dL	>40 mg/dL
C-reactive protein (CRP)	0.68 mg/dL	<0.3 mg/dL

Glucose tolerance test:

Time:		
0 (fasting)		
Glucose	125 mg/dL	<100 mg/dL
Insulin	7.7 μU/mL	<10 μU/mL
½ hour		
Glucose	145 mg/dL	<200 mg/dL
Insulin	78 μU/mL	40–70 μU/mL
1 hour		
Glucose	180 mg/dL	<200 mg/dL
Insulin	99.9 μU/mL	50–90 μU/mL
2 hours		
Glucose	160 mg/dL	<140 mg/dL
Insulin	98 μU/mL	6–50 μU/mL

Normal values from www.inciid.org

Below is an ultrasound of Charlene's left ovary. The image was obtained from the reflection or transmission of ultrasonic waves through the ovary.

The blue circle surrounds the ovary. Numerous small cysts fill the space.

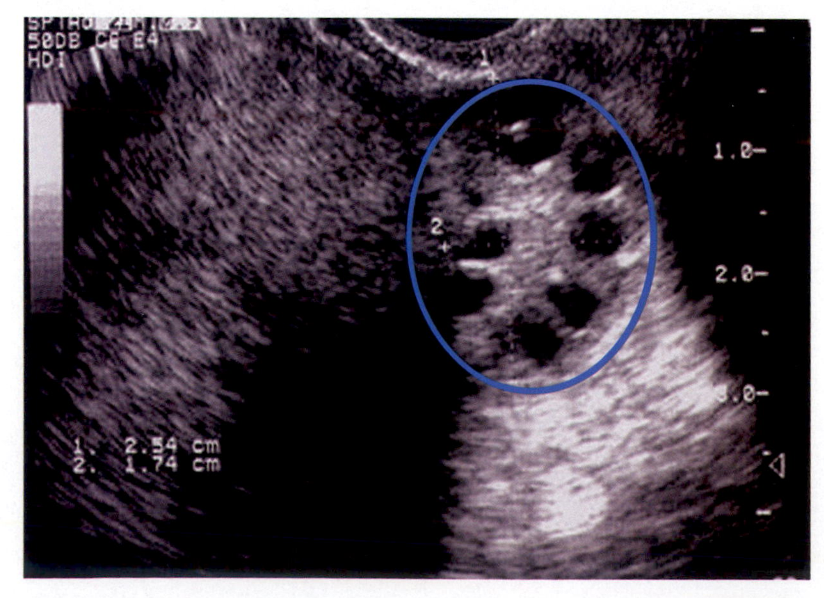

From: Barbieri, RL, Ehrmann, DA. Diagnosis of polycystic ovary syndrome in adults. In: *UpToDate*, Basow, DS (Ed), UpToDate, Waltham, MA 2009. Copyright © 2009 UpToDate, Inc. For more information visit www.uptodate.com

Below is an ultrasound photo of a normal ovary with a mature follicle, mid-cycle.

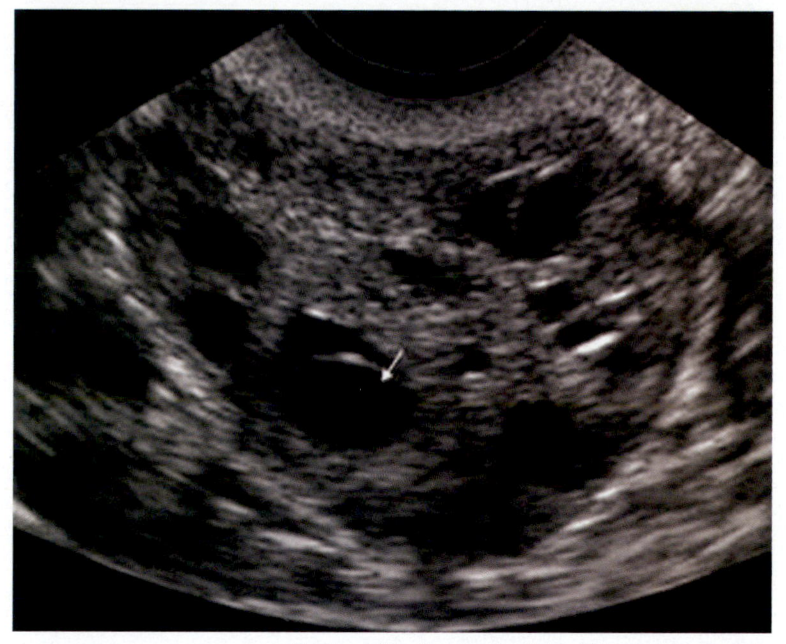

Courtesy of the American Institute of Ultrasound in Medicine

6 What do the results of Charlene's lab tests suggest? (*Hint:* Make a graph of her results and normal values.)

7 How will Charlene be treated?

Answers to Questions

1 *Assess Charlene's height and weight by calculating her body mass index (BMI) using the online calculator at: http://www.nhlbisupport.com/bmi/. Comment on your findings.*

Charlene's BMI is 30.3. A BMI > 22 is indicative of obesity. Charlene is at greater risk for chronic diseases such as hypertension and Type 2 diabetes.

2 *Assess Charlene's blood pressure reading.*

A blood pressure reading of 144/92 in a 13-year-old suggests that she has stage 1 hypertension. Charlene should have three readings taken a week apart to confirm the diagnosis. Her blood pressure should be taken while she is relaxed and has been seated for 5 minutes, with her feet on the floor. A reduced sodium/calorie diet and exercise can bring her blood pressure to normal (120/80 or below), but the specialist also may prescribe an antihypertensive until she is able to change her diet/exercise patterns so she can lose weight.

3 *What problem is causing the hyperpigmentation and hyperkeratosis on Charlene's neck?*

Charlene has Acanthosis nigricans (also known as Acanthosis and Harran's syndrome). It is commonly seen in obese individuals with Type 2 diabetes or a family history of Type 2 diabetes. It is reversible with weight loss and exercise.

4 *What tests would diagnose the underlying causes of this problem?*

Acanthosis nigricans is associated with hyperinsulinemia. A fasting blood glucose level between 100 and 124 signals pre-diabetes and a high risk of developing the disease. If her fasting blood glucose is 125 mg/mL or higher, she will be diagnosed with Type 2 diabetes. A glucose tolerance test in which a standard glucose dose is consumed followed by blood draws at ½ hour, 1 hour, and 2 hours for both glucose and insulin can confirm hyperinsulinemia. Typically, hyperinsulinemia precedes diabetes by decades. In addition to being associated with a greatly elevated risk of Type 2 diabetes, hyperinsulinemia also increases the risk of hypertension and high levels of blood fats called triglycerides and low levels of "healthy heart" lipids called HDL.

5 *What problem is causing Charlene to have irregular periods and pelvic pain, and to grow hair in unusual locations?*

Hyperinsulinemia, obesity, hypertension, Acanthosis, irregular periods, pelvic pain, acne, and excess hair growth on the face, fingers, and toes, called hirsutism, suggest that Charlene has polycystic ovary syndrome (PCOS). Although genetics may play a role, the most common cause of PCOS is obesity with hyperinsulinemia. Weight loss can reverse most, if not all, of the symptoms in adolescents. PCOS and "metabolic syndrome" are related.

Metabolic syndrome involves the following:

- obesity with increased visceral and abdominal fat
- hyperinsulinemia
- insulin resistance and Type 2 diabetes
- hypertension
- increased blood markers of inflammation (e.g., C-reactive protein)

6 *What do the results of Charlene's lab tests suggest?*

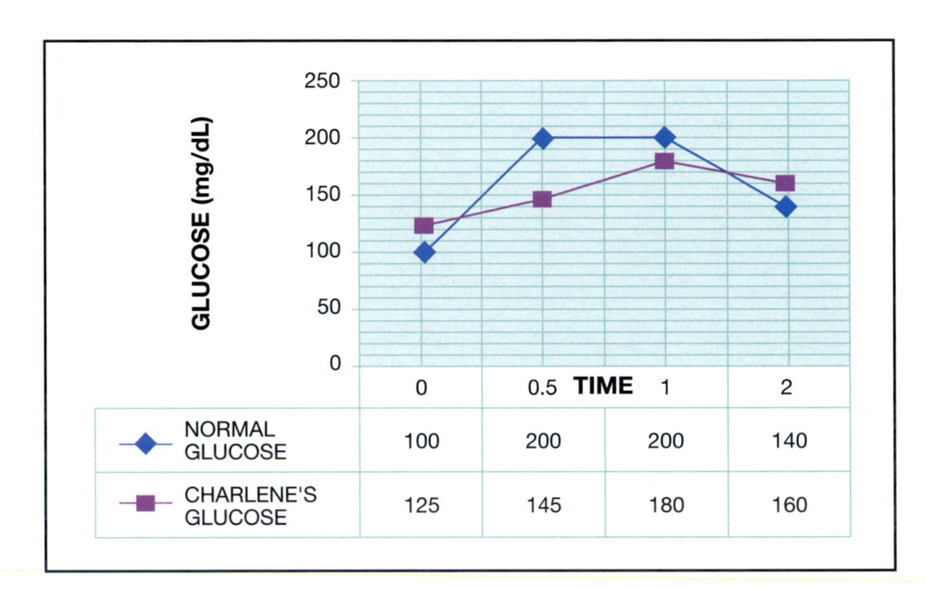

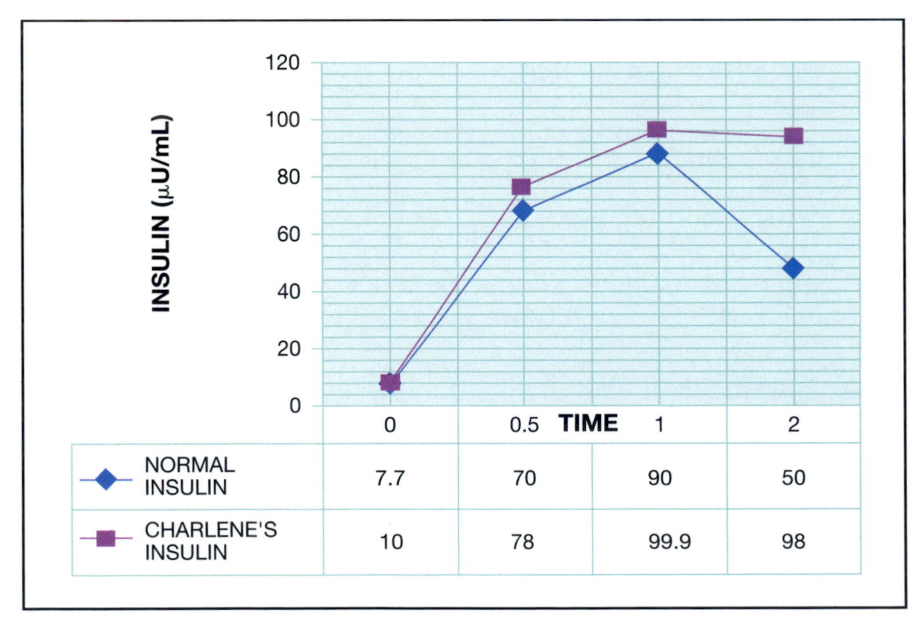

Charlene has hyperinsulinemia, Type 2 diabetes, and polycystic ovary syndrome (PCOS). She also has elevated triglycerides and LDL cholesterol and low HDL cholesterol. These results, taken with her blood pressure reading and BMI suggest that she has metabolic syndrome and PCOS.

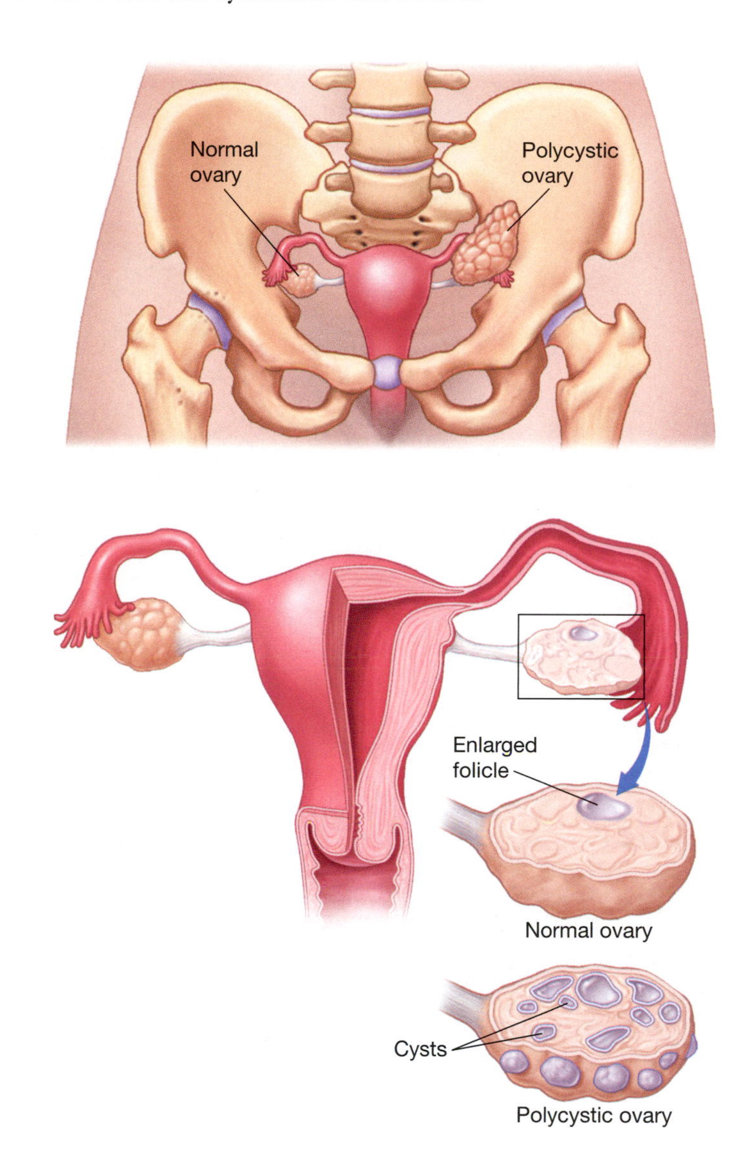

7 *How will Charlene be treated?*

Charlene and her mother were referred to a registered dietitian (RD) for intensive, weekly medical nutrition therapy sessions to increase her fruit, vegetable, and fiber intake and lower her fat and calorie intake. Charlene and her mother also joined a walking club, and they walk for an hour every evening after dinner. On the weekends, they hike or bike in the nearby hills. When weather is too hot, they swim in the local pool. They log 10 hours of exercise each week and at least 10,000 steps a day.

Charlene was started on a low-dose antihypertensive but she was able to discontinue it in 6 months when her blood pressure decreased to 129/79. (Note: The blood pressure goal for individuals with diabetes is <130/80.) After a year, her blood pressure was 115/70. Charlene also was given a drug called metformin to control her blood glucose; this drug also helps to treat the symptoms of PCOS. The doctor also recommended a low-dose birth control pill to help regulate her periods, but Charlene declined this prescription because it is against her religion to use contraceptives. She was prescribed a topical lotion for her acne.

After a year of many, many behavior changes, Charlene is a happy, outgoing, 14-year-old with the lead in the school musical. Her Acanthosis is gone, and her acne and hirsutism are much improved. She has lost 35 pounds (and her mother, 30 pounds). Her periods are still irregular, but her pelvic pain is minimal. Her LDL is 103, and her triglycerides are 150 and HDL 47.

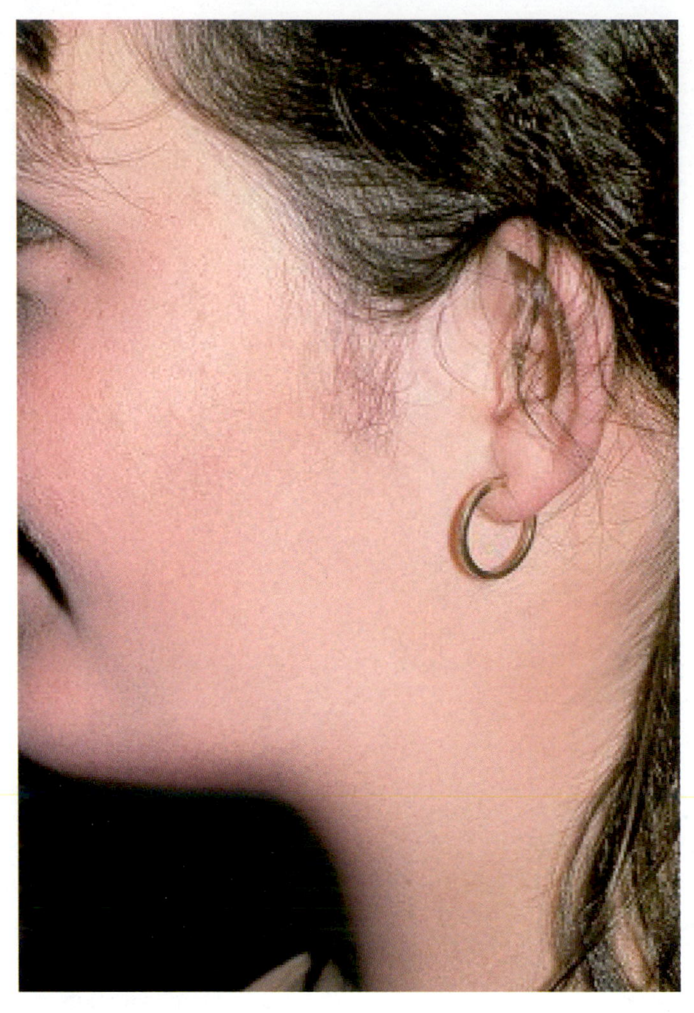

Original image was published in *The Journal of American Academy of Dermatology*, Vol. 55, Lindy P. Fox, *Diseases of the Skin: A Color Atlas and Text*, 2nd Edition, Copyright Elsevier 2006.

References

NIH Body Mass Index calculator
http://www.nhlbisupport.com/bmi/

Diabetes Monitor: Acanthosis nigricans
http://www.diabetesmonitor.com/b313.htm

Acanthosis nigricans. (1999). Montana Diabetes Project. MT Department of Public Health & Human Services
http://www.dphhs.mt.gov/PHSD/Diabetes/pdf/AcanBrochure.pdf

Diabetes Monitor: Polycystic Ovary Syndrome
http://www.diabetesmonitor.com/b219.htm

University of Virginia Health System: Polycystic Ovary Syndrome
http://www.healthsystem.virginia.edu/uvahealth/adult_endocrin/polycyst.cfm

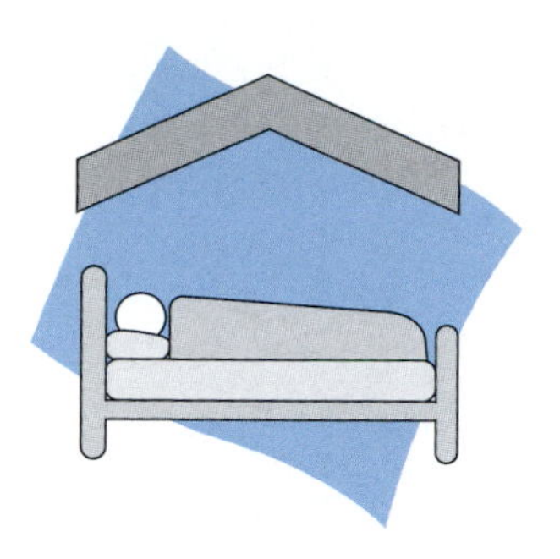

2

Unmade Beds in the B&B

Jane and John (fraternal twins, age 56) and their spouses bought a bed-and-breakfast (B&B) a year ago. The past year was stressful while they secured business loans, renovated the 150-year-old farmstead, and developed a business and marketing plan. Next month will be the grand opening of the 16-room B&B. Today the couples are reviewing the year and anticipating the future.

Jane says, "I'm exhausted—as I have been for the past month. I know my job is to get the beds ready, but I'm so fatigued that it takes me half an hour to get each bed changed. That's quite a workout. I get out of breath! Now my back is bothering me and I have an appointment in an hour for an MRI at the hospital."

At this, John interrupts and says "I'm coming with you. I have this sharp pain running from my left shoulder up my neck and down my arm. I think I'll get checked out at the emergency department. All this painting is taking its toll."

John and Jane both have ECGs and blood tests and get MRIs.

John's ECG

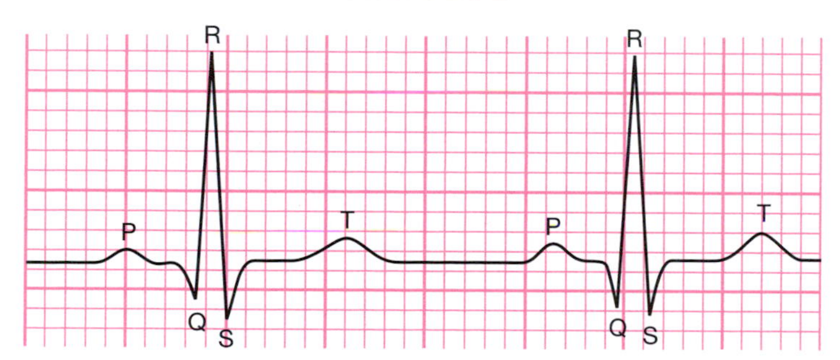

Jane's ECG

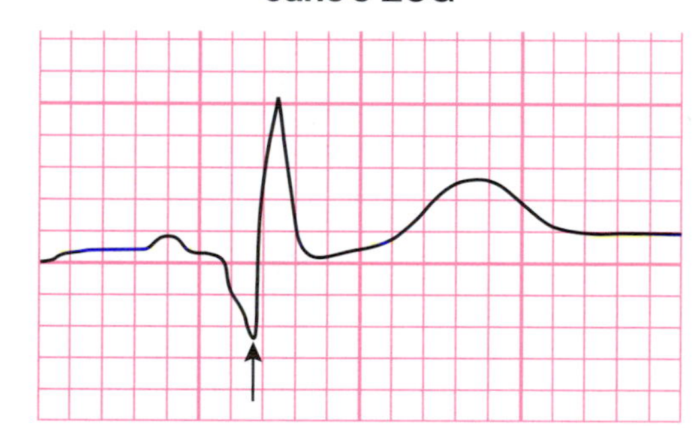

John's Selected Lab Values

		Normal Range
Creatine phosphokinase (CK)	98 units/L	55–170 units/L
CK-MB/CK isoenzyme	0%	0% of total CK
Myoglobin	27mcg/L	<90 mcg/L
Troponin T (ng/mL)	Not detected	<0.2 ng/mL

Jane's Selected Lab Values

		Normal Range
Creatine phosphokinase (CK)	148 units/L	30–135 units/L
CK-MB/CK isoenzyme	4%	0% of total CK
Myoglobin	30mcg/L	<90 mcg/L
Troponin T	3.2 ng/mL	<0.2 ng/mL

An MRI (magnetic resonance image) is a diagnostic imaging technique whereby the patient is put in a strong, uniform magnetic field. Protons absorb energy from the magnetic field and then emit radio waves as their excitation decays. These radiofrequency signals are converted into three-dimensional images. MRIs do not expose patients to ionizing radiation.

Normal MRI of Shoulder

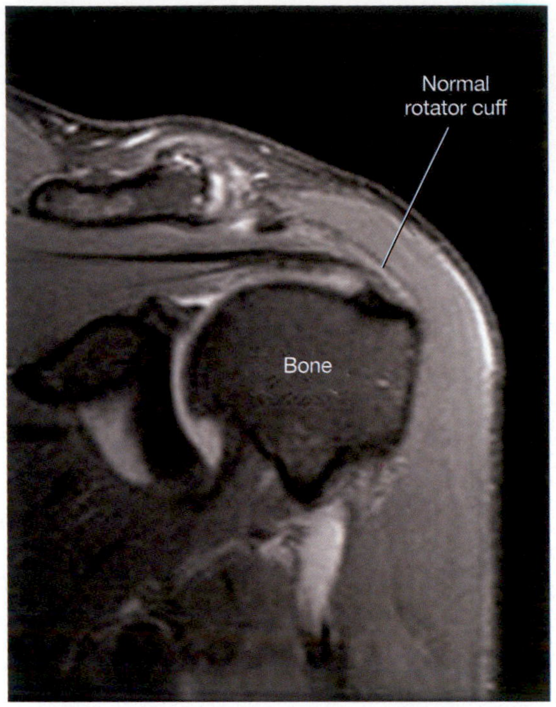

Photo courtesy of Intermountain Medical Imaging, Boise, Idaho

John's MRI of His Shoulder

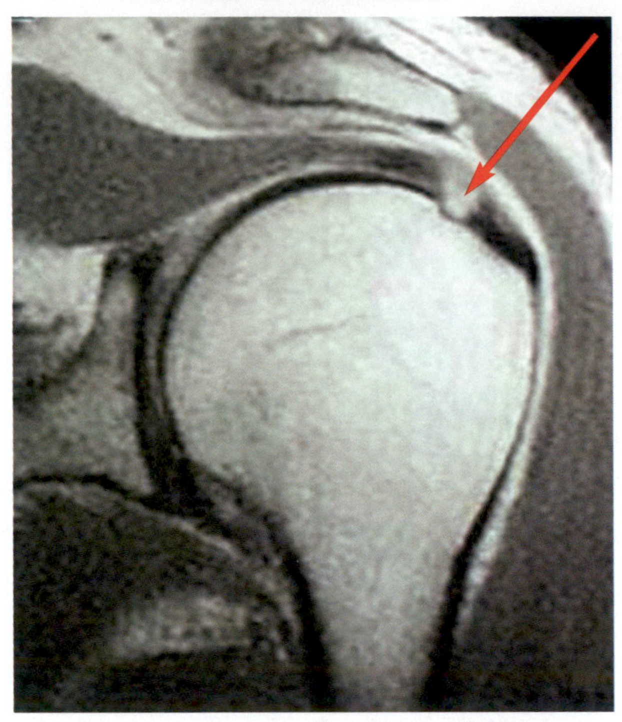

Photo courtesy of Emedx.com

Normal MRI of Spine

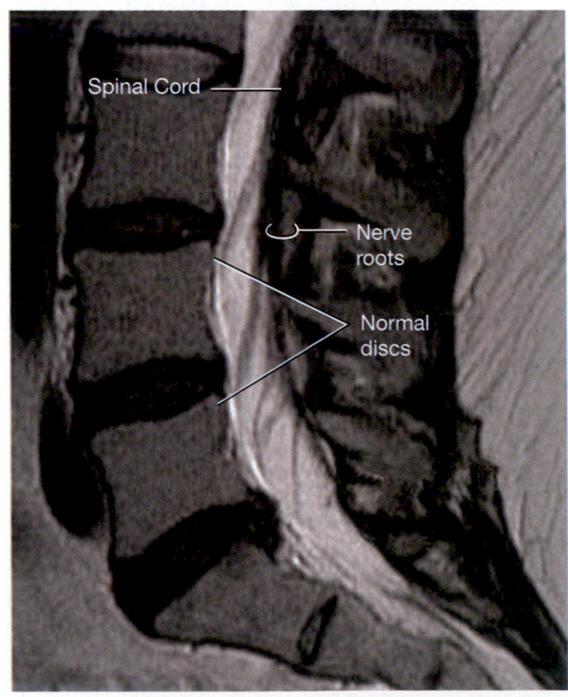

Photo courtesy of Intermountain Medical Imaging, Boise, Idaho

Jane's MRI of Her Spine (S1 degeneration)

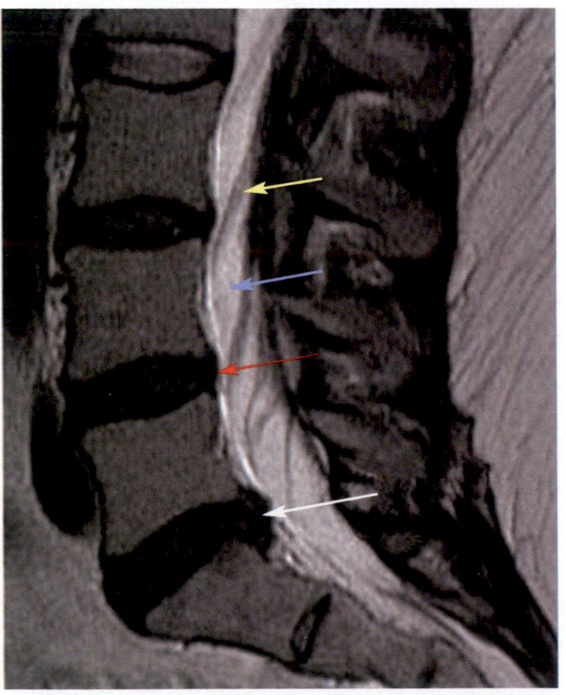

Photo courtesy of Intermountain Medical Imaging, Boise, Idaho

Questions

Name ______________________ Section ________

1 What problem can these selected lab values detect? What is the significance of each lab value?

2 What do John's ECG and lab values suggest?

3 What do Jane's ECG and lab values suggest?

4 What does the red arrow on John's MRI suggest?

5 What do the regions indicated by the red, blue, yellow, and white arrows on Jane's MRI suggest?

6 What are gender-related differences in John's and Jane's symptoms?

7 What treatments/medications will John need?

8 What treatments/medications will Jane need?

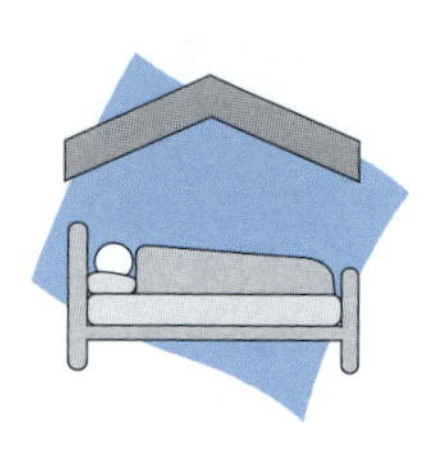

Answers to Questions

1 *What problem can these selected lab values detect? What is the significance of each lab value?*

These lab tests can indicate the presence of an ongoing or previous myocardial infarction (acute MI, or heart attack). When myocardial cells are deprived of oxygen during an MI, the cells are injured and release enzymes and other proteins into the blood. Myoglobin (Myo) rises early in an MI and goes back to normal within 2 days. An elevated myoglobin level suggests that an MI is currently in progress. Cardiac troponin T (Tnt) levels peak 3–4 days after the onset of an MI. Elevated troponin T levels can indicate that an MI took place in the previous 1–6 days. Troponin T levels correlate with the extent of myocardial damage.

Creatine phosphokinase (CK) is an enzyme present in all muscle cells. It uses ATP to synthesize phosphocreatine. Muscle cells use phosphocreatine as a rapid-energy reserve of ATP for biochemical reactions. CK becomes elevated, *and* CK from the heart (CK-MB) is detected 1–2 days post-MI.

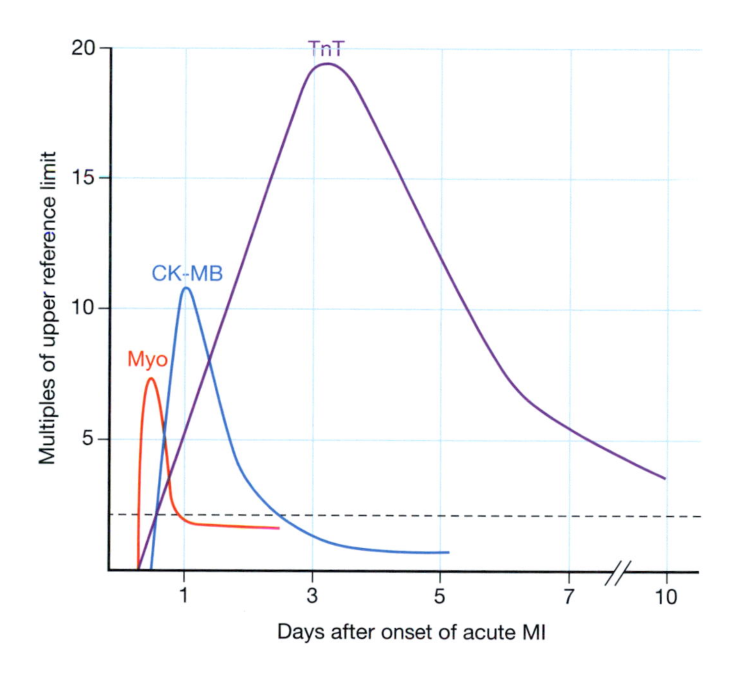

Cardiac enzyme levels following onset of acute myocardial infarction: Myo = Myoglobin; Tnt = Cardiac troponin T; CK-MB = Creatine phosphokinase (cardiac)

2 *What do John's ECG and lab values suggest?*

Although John was having symptoms of a possible MI, his lab values and ECG suggest that he did not have a heart attack. His lab values for creatine phosphokinase, CK-BM/CK isoenzyme, myoglobin, and troponin are all normal. His ECG has no Q-wave or T-wave inversion, and the amplitude of his R wave is normal.

3 *What do Jane's ECG and lab values suggest?*

Jane's ECG (repeated below) is abnormal. She has a pathological Q wave and diminished amplitude of her R wave because of cardiac muscle tissue necrosis (cell death). She does not have T-wave inversion. T-wave inversion suggests current ischemia. T-waves typically return to normal by about 4 weeks after an MI.

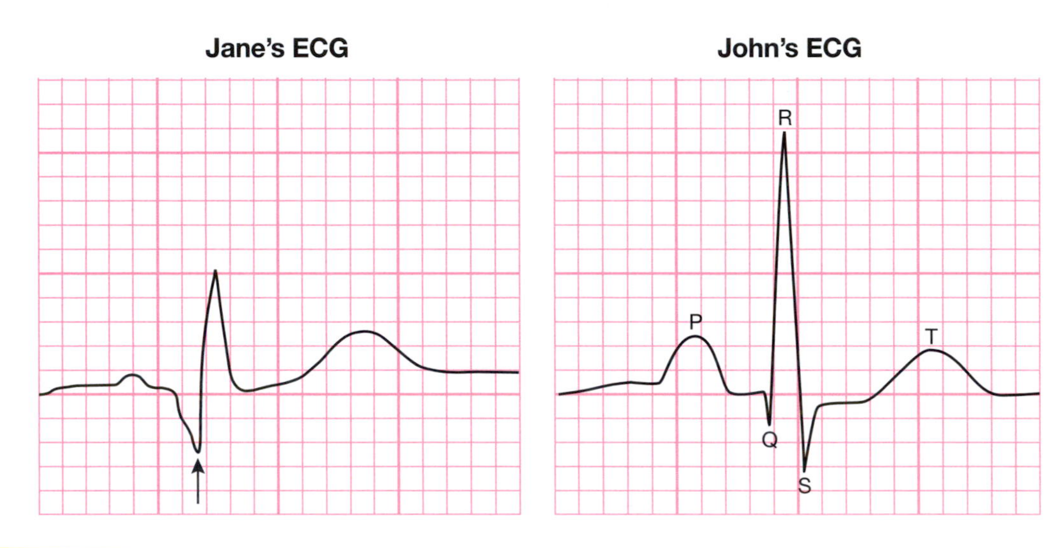

Jane's lab values (repeated below) suggest that she has had an MI. Her CK-MB is 4% of the total CK value. Her troponin T level is 20 times the upper normal limit. Her myoglobin level is normal. Taken together, these lab values suggest that Jane had an MI 1–4 days ago.

Jane's Selected Lab Values

		Normal Range
Creatine phosphokinase (CK)	148 units/L	30–135 units/L
CK-MB/CK isoenzyme	4%	0% of total CK
Myoglobin	30mcg/L	<90 mcg/L
Troponin T	3.2 ng/mL	<0.2 ng/mL

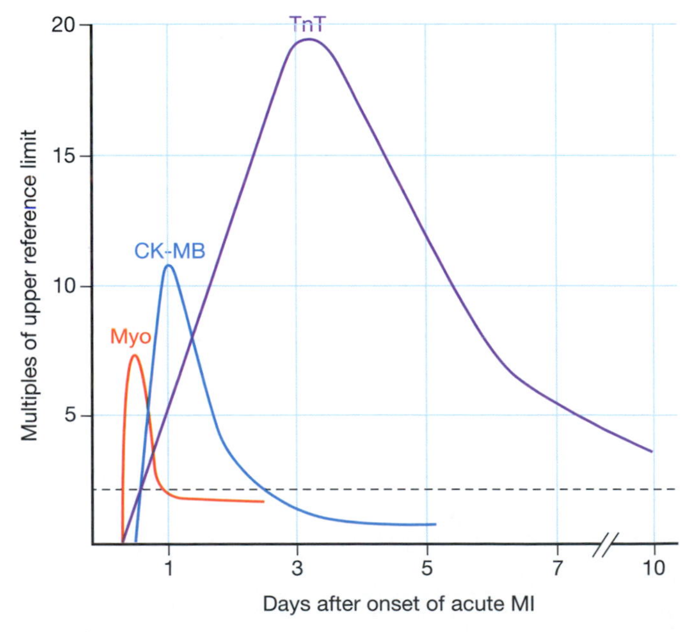

Diagram courtesy of Roche diagnostics http://www.roche.com/home.html

4 *What does the red arrow on John's MRI suggest?*

There is fluid in John's rotator cuff, suggesting a tear in his rotator cuff.

5 *What do the regions indicated by the blue, yellow, red, and white arrows on Jane's MRI suggest?*

The blue arrow points to a normal spinal canal. The yellow arrow points to a nerve root. The red arrow points to a normal disc. The white arrow points to a herniated disc at S1.

6 *What are gender-related differences in John's and Jane's symptoms?*

The symptoms of a torn rotator cuff and herniated disc at S1 are identical in men and women. The symptoms of MI, however, may differ. Men often have what are known as "classic" symptoms of pain radiating from the chest to the left neck area and down the left arm. Pain typically lasts at least 15 minutes and is not relieved by rest. Women may have these symptoms also, but often women have more vague symptoms. This may be because often the smaller coronary blood vessels rather than the coronary arteries occlude in women. Women may have back pain or symptoms of epigastric pain that is mistaken for indigestion. The fatigue and breathlessness that Jane experienced may be the only symptoms. Women also are more likely than men to experience nausea and vomiting or jaw pain.

7 *What treatments/medications will John need?*

John will need a surgical repair of his rotator cuff, followed by physical therapy.

8 *What treatments/medications will Jane need?*

Jane will require further assessment of her cardiovascular profile: blood pressure, LDL and HDL cholesterol, triglycerides, cReactive protein (CRP, a measure of inflammation), and homocysteine. She will probably undergo

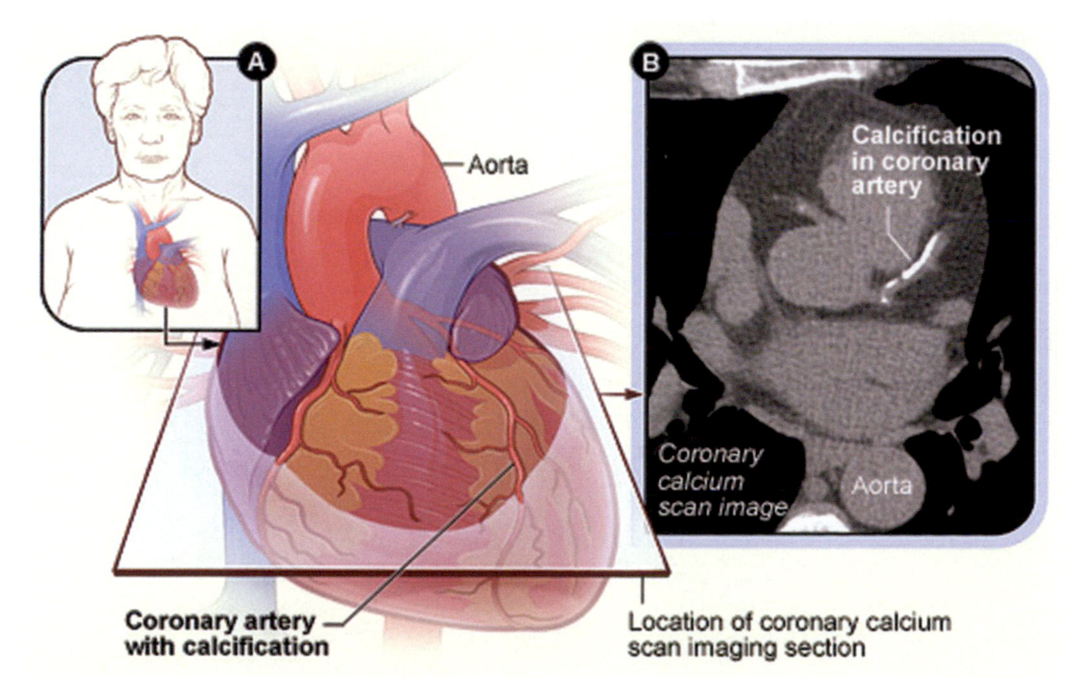

Courtesy of The National Heart, Lung and Blood Institute (NHLBI)

a cardiac catheterization to determine whether she needs a stent to open an occluded coronary artery. She also may be given a cardiac calcium scan to detect small-vessel damage in the heart. Jane probably will be assessed for anxiety and depression because, if untreated, these conditions are known to increase the risk of a second MI.

She probably will be prescribed medications that have been shown to reduce the risk of a second MI. These may include a beta blocker and an angiotension converting enzyme inhibitor (ACE inhibitor) to control blood pressure, low-dose aspirin or Plavix to reduce platelet aggregation, and a stool softener to help her avoid constipation (the Valsalva effect from straining during defecation can precipitate an MI). She will be referred to a cardiac rehabilitation program to increase her exercise tolerance and to a stress-reduction program.

Jane will be encouraged to follow lifestyle changes including increased physical activity (60–90 minutes of moderate-intensity activity such as brisk walking daily), alcohol moderation, sodium restriction, and a diet rich in fresh fruits and vegetables and low in saturated fat with two servings of oily fish weekly. Read more at:

http://www.americanheart.org/presenter.jhtml?identifier=1200000

The effect of specific risk factors on risk of MI can be determined on the following website:

http://www.goredforwomen.com/hcu/index.aspx

Jane also will be given physical therapy and exercises to help reduce pain from her degenerated disc.

References

Go Red for Women
http://www.goredforwomen.com/hcu/index.aspx

The American Heart Association
http://www.americanheart.org/presenter.jhtml?identifier=1200000

What is a Coronary Calcium Scan? National Heart, Lung and Blood Institute
http://www.nhlbi.nih.gov/health/dci/Diseases/cscan/cscan_whatis.html

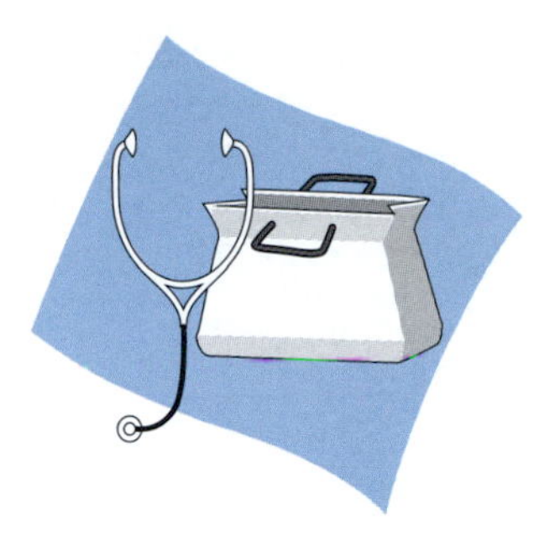

3

"My Heart is in My Stomach!"

Mrs. V., age 52, comes into the hospital walk-in clinic and says, "My 6-year-old granddaughter got a medic kit for her birthday. She was using the stethoscope to hear my heartbeat, and she couldn't find it— it *is* a toy after all. But then she put the stethoscope on my stomach and said she heard a loud heartbeat. I can feel it too, just above my navel. Is it possible that my heart is in the wrong place?

My granddaughter is telling everyone, "Grandma has a baby in her tummy just like mommy! You can feel the heartbeat!"

The notes from the triage nurse indicate the following:

Female, age 52
BP: 176/94
Height (self-reported): 5'2"
Weight (self-reported): 165 lbs.
Smoker: 1–2 packs/day for 33 years
Heart rate: 88 with occasional PVCs (1 per minute)
Last period was 3 months ago
Pulsing mass above umbilicus, upper left of midline, palpable when supine.

A physician's assistant (PA) sees Mrs. V. and takes a pregnancy test, then orders an X-ray.

Normal X-ray

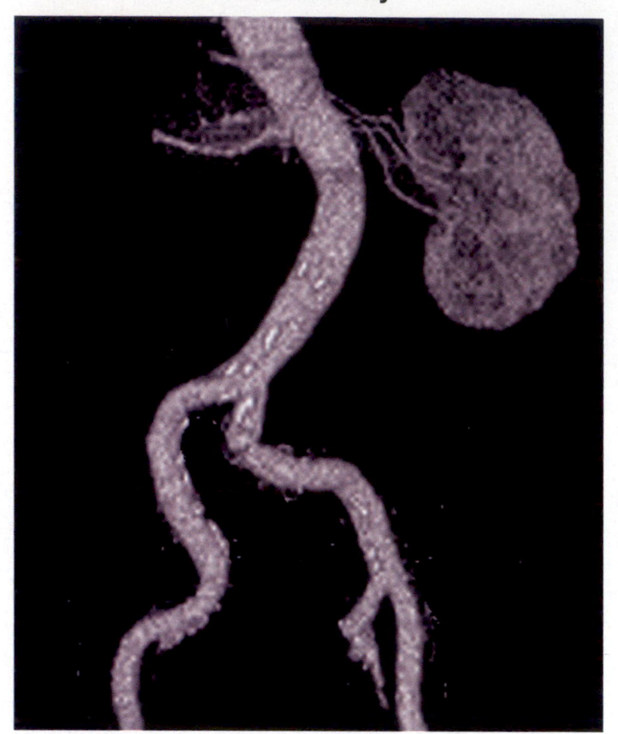

Original image courtesy of www.pennhealth.com

Mrs. V.'s X-ray

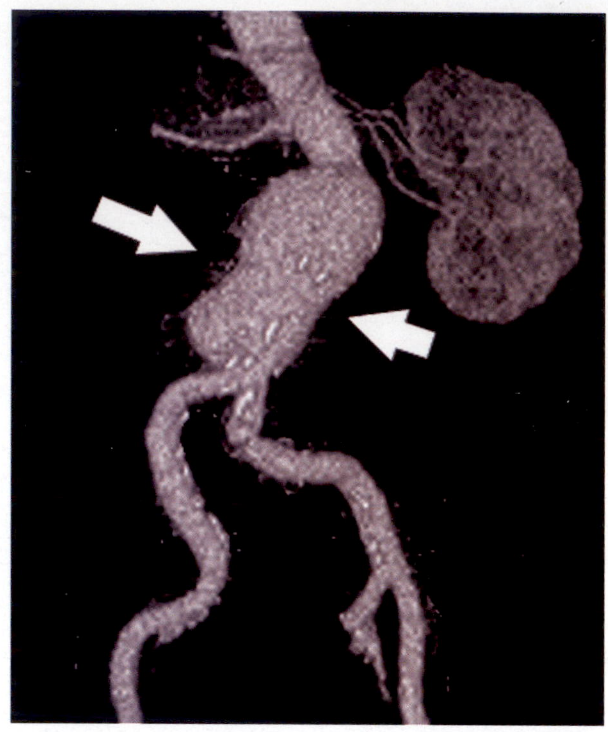

Courtesy of www.pennhealth.com

A Multi Row CT Angiogram Scan is then ordered for Mrs. V. This type of CT scan has multiple rows of detectors. Through a process called volume rendering, the final image appears to be three-dimensional.

Normal Scan

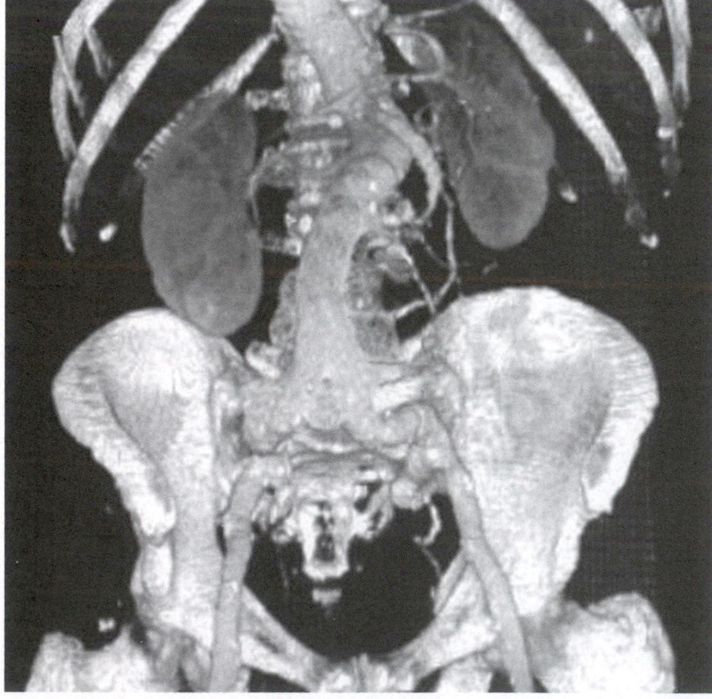

Original image courtesy of Madison Radiologists: http://www.madisonradiologists.com/

Mrs. V.'s CT Scan

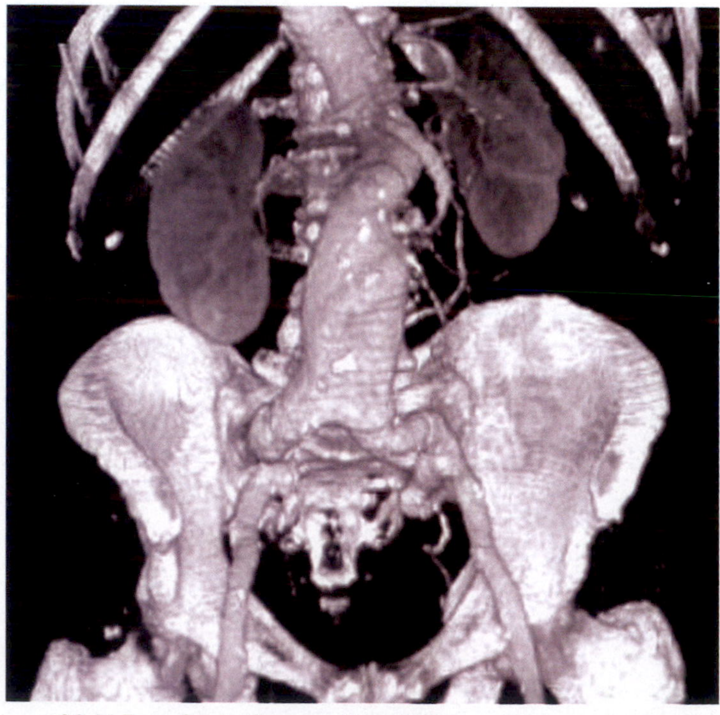

Multi Row CT Angiogram Scan that is volume-rendered.

Courtesy of Madison Radiologists: http://www.madisonradiologists.com/

Questions

Name ______________________________ Section ________

1 Why was a pregnancy test ordered for Mrs. V.?

2 What does Mrs. V.'s X-ray show? (Hint: What do the arrows point to?)

3 What does Mrs. V.'s CT scan show?

4 What does Mrs. V. have? What probably caused this to happen?

5 If Mrs. V. does nothing about this, what might happen to her?

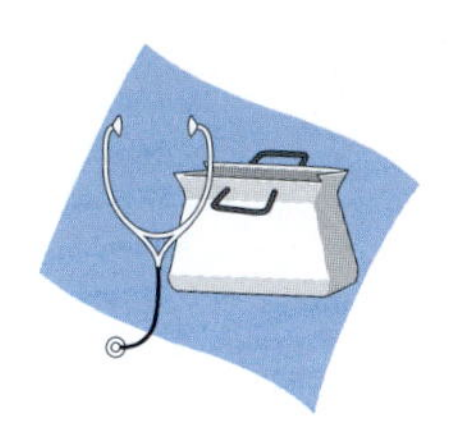

Answers to Questions

1 *Why was a pregnancy test ordered for Mrs. V.?*

Mrs. V. is still having periods, so she might be pregnant. While early (3 months' gestation) pregnancy would not cause her symptoms, she might have had an earlier date of conception and continued to have some periodic bleeding. Fetuses become active beginning in the second trimester and can be felt as a "flutter." Hiccups are common in some fetuses and can cause a sensation of a "pulsing mass." If Mrs. V. is pregnant, an X-ray and certain medications would endanger the fetus.

2 *What does Mrs. V.'s X-ray show?*

The X-ray shows an area of calcification on an enlarged section of her abdominal aorta.

3 *What does Mrs. V.'s CT scan show?*

The CT scan confirms an enlargement of the abdominal aorta. The size of the enlargement can be calculated from the CT scan.

4 *What does Mrs. V. have? What probably caused this to happen?*

Mrs. V. has an abdominal aortic aneurysm (AAA). Abdominal aortic aneurysms occur when a section of the aorta weakens and bulges out like a stretched balloon. AAAs are more common in older adults (over age 60) and in men. Mrs. V. may have a genetic predisposition. More likely, it developed from atherosclerosis from long-term untreated high LDL cholesterol and hypertension. In men, chronic consumption of two or more alcoholic drinks per day is a risk factor. Often, people (especially women) have no symptoms and it is identified from an X-ray taken for another problem. Sometimes men complain of groin pain. Large aneurysms (> 4–6 cm) may cause mild to moderate back pain. The U.S. Preventive Services Task Force now recommends that men (but not women) aged 65–75 who have ever smoked get a one-time ultrasound screening for AAAs.

5 *If Mrs. V. does nothing about this, what might happen to her?*

Abdominal aortic aneurysms can rupture with a mortality rate that ranges from 50%–90%.

If the aneurysm is small (less than 5 cm; the normal diameter of the aorta is 2 cm), her cardiologist may recommend "watch and wait" by tracking the size of the aneurysm every 3 months. Treatment for hypertension and elevated cholesterol will be initiated. She *must* stop smoking and she should avoid heavy lifting (such as grandchildren!). If the aneurysm is larger than 5 cm or growing, surgery is required to avoid rupture. A mesh stent may be implanted in the artery to stabilize it and prevent it from growing larger. A ruptured ("dissected") abdominal aortic aneurysm may be repaired with a graft; however, the procedure has a high risk of mortality.

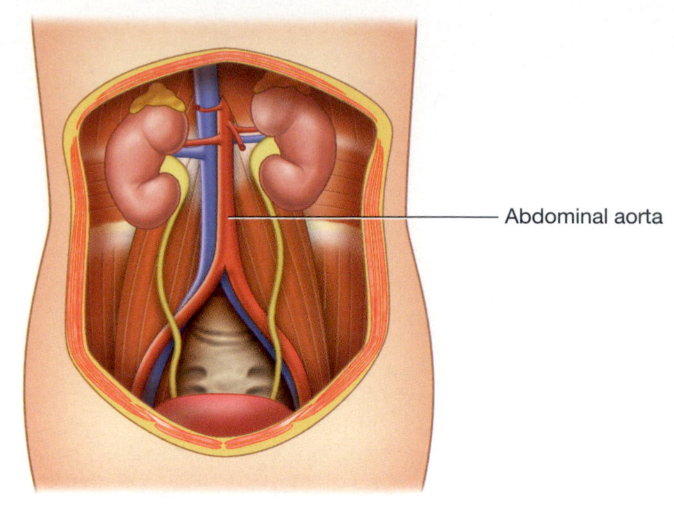

Normal Anatomy

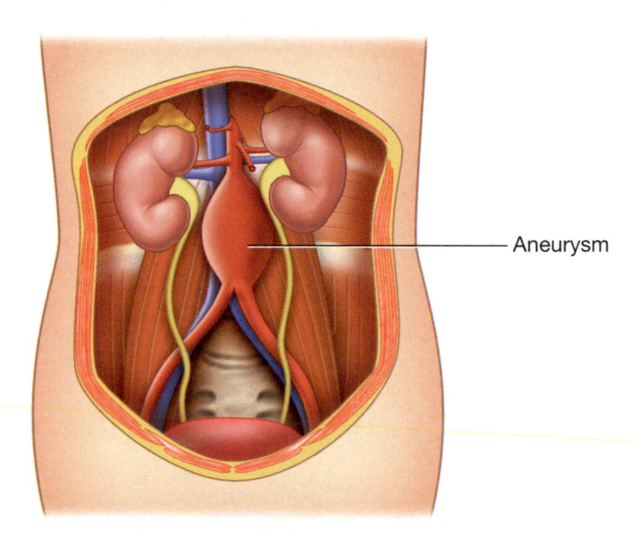

Abdominal aortic aneurysm

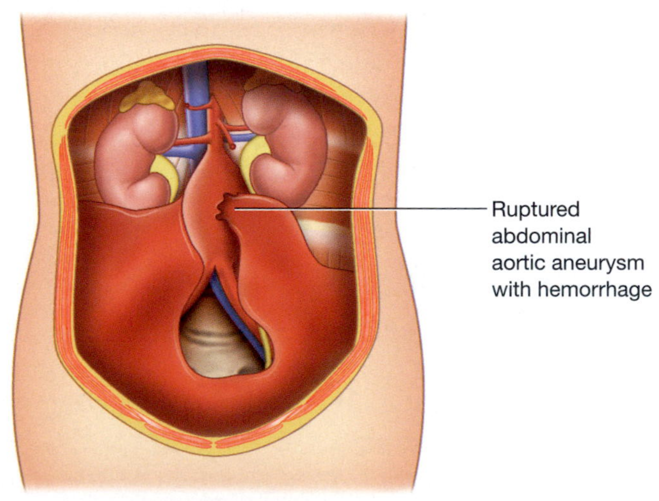

Eventual condition

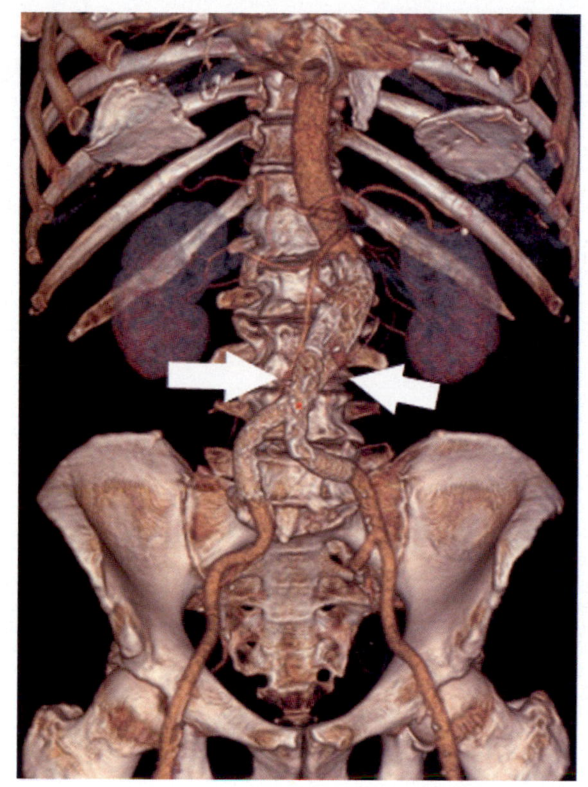

Courtesy of www.pennhealth.com

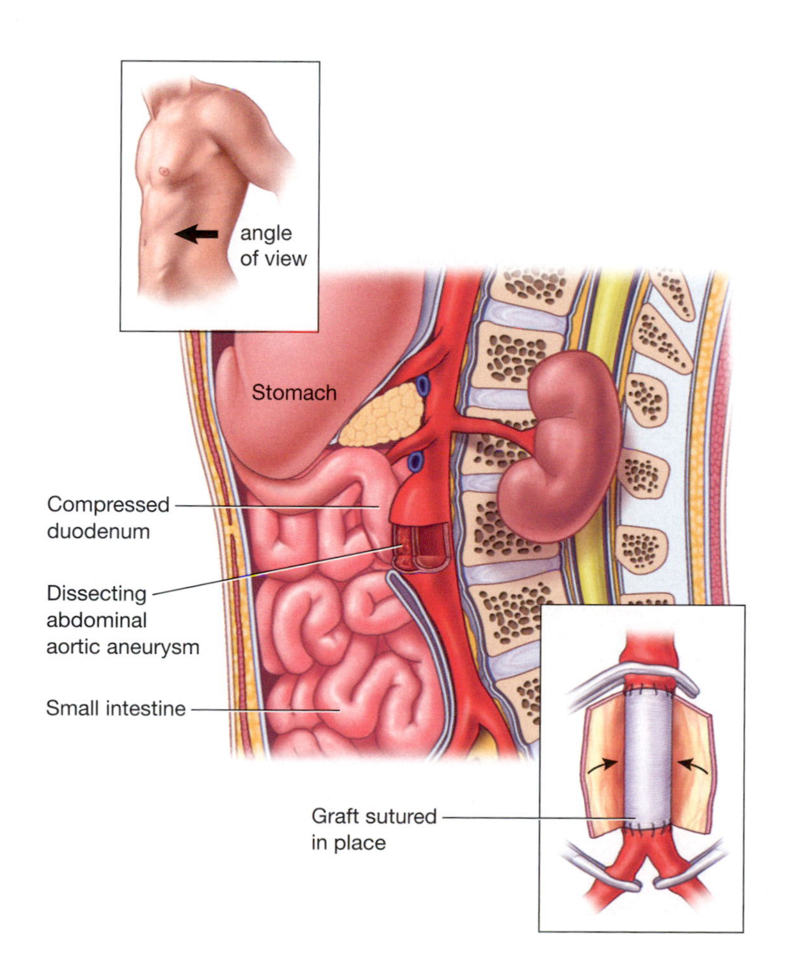

References

Abdominal Aortic Aneurysm. University of Pennsylvania
http://www.pennhealth.com/int_rad/health_info/aaa.html

Screening: Abdominal Aortic Aneurysm. U.S. Department of Health & Human Services, Agency for Healthcare Research and Quality (AHRQ)
http://www.ahrq.gov/clinic/uspstf/uspsaneu.htm

4

"My Professor Makes No Sense!"

Sophomore Susan Jones is in the Student Advisory Center meeting with her tutor. "I don't know what else to do. I go to every class. I read the book. I take careful notes. I even tried taping the class like you suggested. I still got a D on the last exam. My professor makes no sense! Everyone says so. Here—listen to the tape."

The tutor clicks on the tape, thinking that this would be a teachable moment to help Susan with active listening and note-taking skills. The professor starts: ". . . inflammatory mediators released immediately from mast cells include histamine, neutrophil chemotactic factor, and eosinophil chemotatic factor of anaphylaxis (known as ECF-A)."

"Well this is difficult material, says the tutor."

"I know," says Susan, but go on with the tape!"

The professor on the tape continues "And sero, seri, Suzie................uh................platelets........uh so the cells go and do it.......uh okay class, you can go now."

"He often goes like that," says Susan. We've lost so much of class this semester with his sort of rambling and losing track of where he is. Does he have Alzheimer's or something? He should retire!"

With concern, the tutor makes a call to the professor's department and finds out that the professor had been hospitalized earlier that day.

Meanwhile, the professor is having tests at the hospital emergency department. Following is what is known about the professor:

Male, age 56

Reports that he has been having episodes for the past two months where he forgets what he is about to say during a lecture and cannot find the words. His vision dims. Episodes typically last about 2 minutes. He thought he was having either anxiety or orthostatic hypotension from his Zestril (lisinopril) prescribed for hypertension so he stopped taking it. Today, his episode lasted almost 10 minutes. He has not been taking his Lipitor (atorvastatin) prescribed for high cholesterol over the past month because he says, "I've been following my diabetic diet" for pre-Type 2 diabetes mellitus diagnosed 6 months ago. He takes Inderal (propranolol, purchased over the Internet) every morning before class to reduce his anxiety while speaking.

Smoker: 1 pack/day
Height: 5'11"
Weight: 195 lbs

	Normal Values
Glucose (4 hrs post-prandial): 165 mg/dL	< 140 mg/dL
HbA1c: 10.4%	2%–5%
Blood pressure: 156/94	120/80
Resting pulse: 52 bpm	60–90 (lower in athletes)
LDL cholesterol: 175 mg/dL	60–130 (ideal is < 100 mg/dL)

Below is a Doppler ultrasound image of the professor's left carotid artery. The image is obtained from the reflection or transmission of ultrasonic waves through the carotid artery.

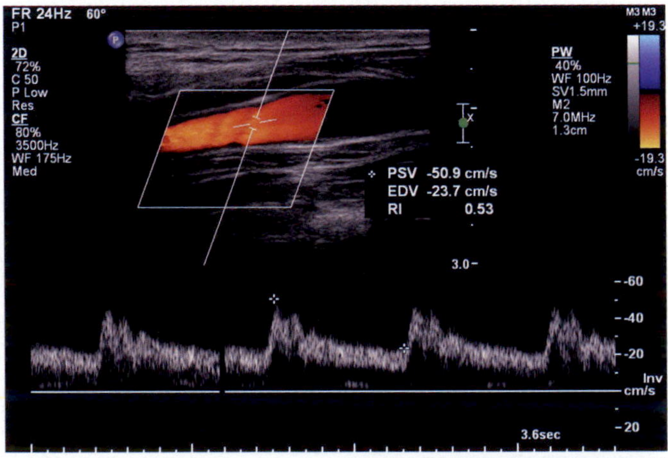

Courtesy of Dr. Mustapha Azzam 2008

Below is a Doppler ultrasound image of the professor's right carotid artery.

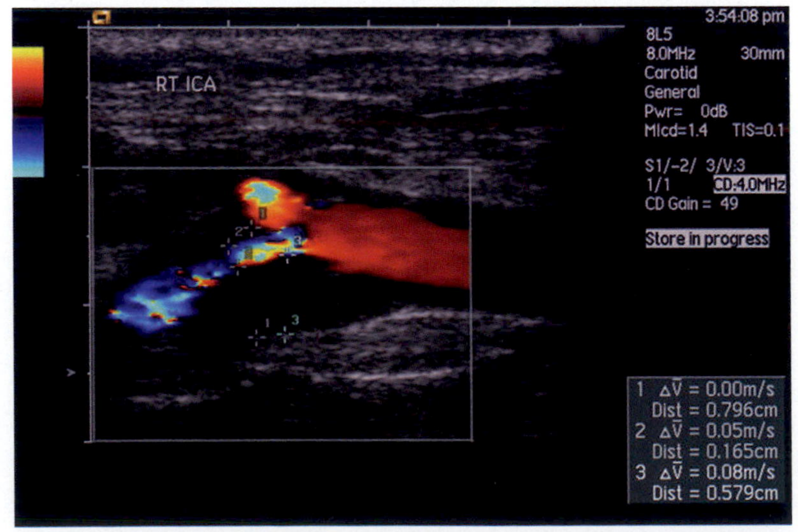

From *A 75-year-old Woman with a Hemispheric Stroke*, Kakkos SK, Geroulakous G, *PLOS Medicine* Vol 2, No 4, e79doi:10.1371/journal.pmed.0020079

Below, right, is an image of an angiogram of the professor's carotid arteries. A catheter was inserted into his femoral artery and moved to the carotid artery. A radio opaque contrast material then was injected. The contrast material blocks the passage of X-rays and allows the visualization of carotid arteries on a fluoroscope (an X-ray machine that projects images on a television monitor).

Left, Right Carotid Artery

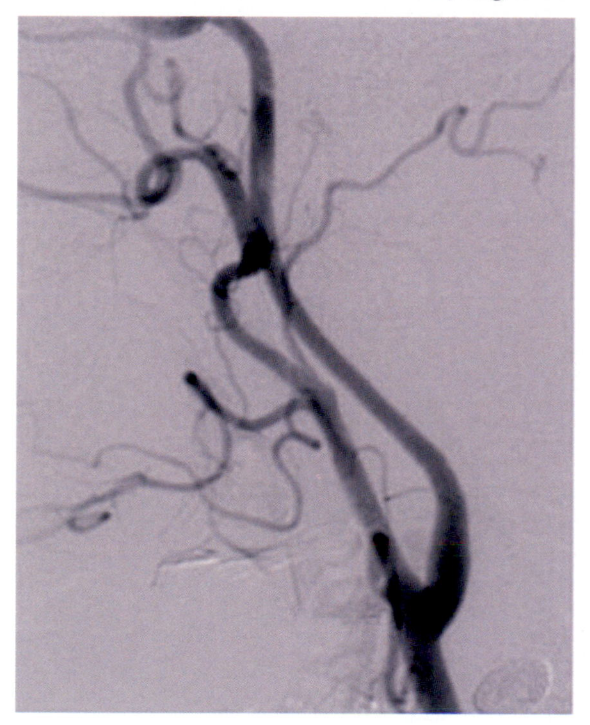

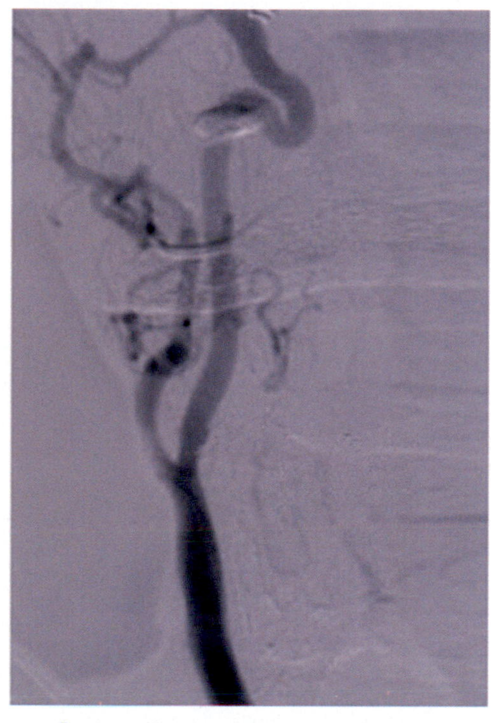

Courtesy of Intermountain Medical Imaging, Boise, ID

Questions

Name ____________________ Section __________

1 What problem caused the professor's symptoms?

2 What do the glucose and HbA1c values suggest?

3 What signs are revealed by the tape-recording, and what do they indicate? How high is the professor's risk for a cerebral vascular accident (CVA, or stroke)?

4 How will the professor's problems be treated?

5 Why does the professor have a slow pulse of 52?

Answers to Questions

1 *What problem caused the professor's symptoms?*

The professor has an occluded right carotid artery caused by a build-up of atherosclerotic plaque. Blood flow to the brain has been reduced intermittently causing his symptoms of a transient ischemic attack (TIA). His uncontrolled LDL cholesterol, Type 2 diabetes mellitus, and hypertension all greatly contributed to his atherosclerosis.

Signs and symptoms of transient ischemic attacks may include sudden onset and sudden disappearance of any of the following:

- slurred or garbled speech
- inability to understand what others are saying
- loss of vision in one or both eyes or double vision
- dizziness or loss of balance and/or coordination
- numbness, weakness, or paralysis in one side of the face and/or one arm or leg

Although TIAs have symptoms similar to symptoms of a stroke (cerebral vascular accident, or CVA), TIAs are characterized by temporary symptoms that disappear within 24 hours.

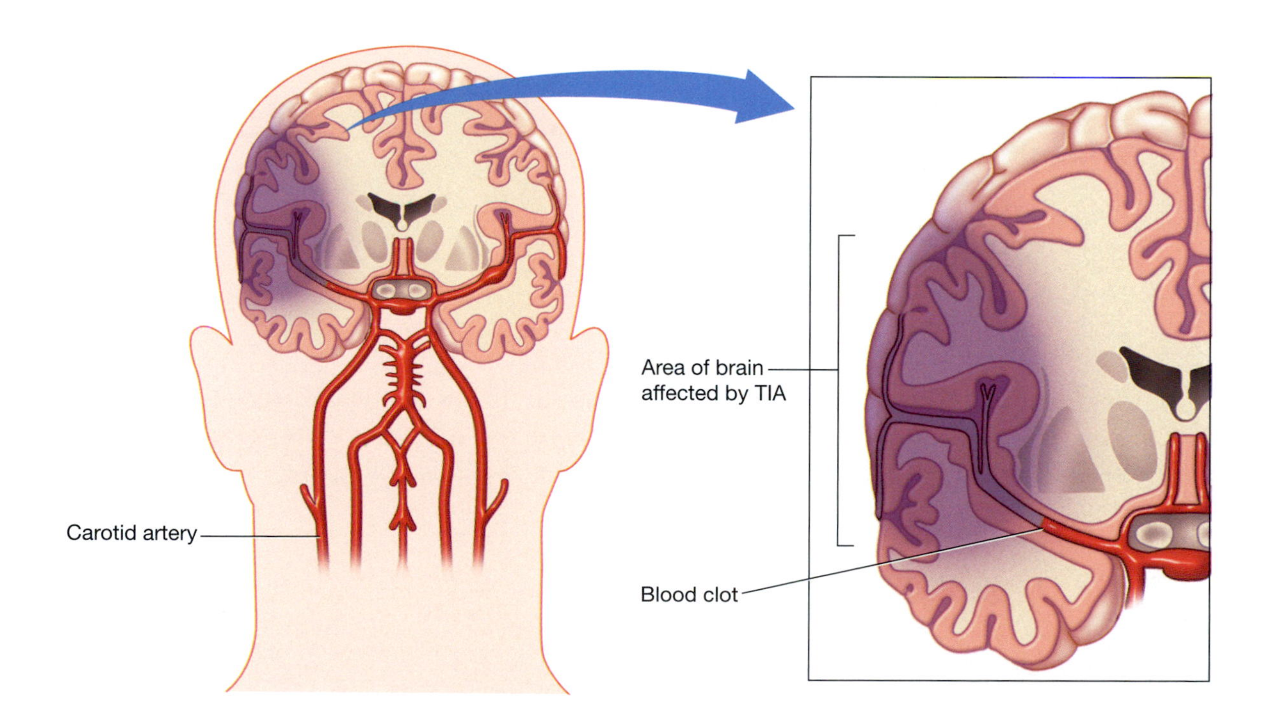

2 *What do the glucose and HbA1c values suggest?*

HbA1c is a measure of percent of hemoglobin that is glycated (bound to glucose). Values above 7% (normal is 5% or less) suggest a chronic elevation in blood glucose over the previous 2–3 months (the lifespan of an average red blood cell). The professor's value of 10.4% indicates that his blood glucose has been poorly controlled over the past 2 months or more. Poorly controlled Type 2 diabetes mellitus greatly increases the progression of atherosclerosis and the risk of both TIA and stroke.

3 *What signs are revealed by the tape-recording, and what do they indicate? How high is the professor's risk for a cerebral vascular accident (CVA, or stroke)?*

The tape suggests the professor has speech impairment without weakness. This indicates that he had a TIA.

The ABCD^2 prognostic score (Johston et al., 2007) predicts the risk of a CVA over the 2-day period following the TIA. The professor's score is calculated below:

ABCD^2 Parameter	The Professor's Score
A (age):	
age ≥ 60 [1 point]	0
B (blood pressure):	
blood pressure ≥ 140/90 mm Hg [1 point]	1
C (clinical features):	
unilateral weakness [2 points]	0
speech impairment without weakness [1 point]	1
D (duration of symptoms):	
duration: >60 minutes [2 points]	0
10–59 minutes [1 point]	1
D (diabetes):	
diabetes [1 point]	1
Total possible score: 6	4

Risk of CVA within 2 days

High: Score = 6–7
Moderate: Score = 4–5
Low: Score = 0–3

Overall, approximately a third of people experiencing a single TIA will have a stroke within the following year. The ABCD^2 score suggests that the professor has a moderate risk of having a stroke within the next two days. Therefore, he will be treated aggressively to reduce his immediate risk of stroke.

4 *How will the professor's problems be treated?*

The professor has a moderate risk of having a stroke over the next two days. He will have a carotid endarterectomy to remove atherosclerotic plaque from his right carotid artery.

The professor will be prescribed oral medication to control his Type 2 diabetes mellitus. He will be prescribed an antiplatelet agent such as low-dose aspirin or Plavix (clopidogrel) to reduce his risk of stroke. He will be advised to quit smoking and adhere to his antihypertensive medication and his "statin" drug for his high LDL cholesterol. The American Heart Association and American Diabetes Association now recommend that individuals with Type 2 diabetes maintain an LDL cholesterol less than 70 mg/ml (as compared to less than 100 mg/ml for patients without diabetes) to control their greatly increased risk of heart disease from Type 2 diabetes. The blood pressure goal for a person with diabetes is now < 130/80 mm Hg.

After full recovery from surgery, the professor will undergo diagnostic testing for coronary artery atherosclerosis (exercise stress testing, angiograms, etc.)

The professor will be referred to a clinical dietitian for nutrition counseling for obesity, hypertension, atherosclerosis, and Type 2 diabetes. His blood pressure, blood glucose, and HbA1c will be monitored closely. He also will be referred to cardiac rehabilitation center for exercise training and monitoring.

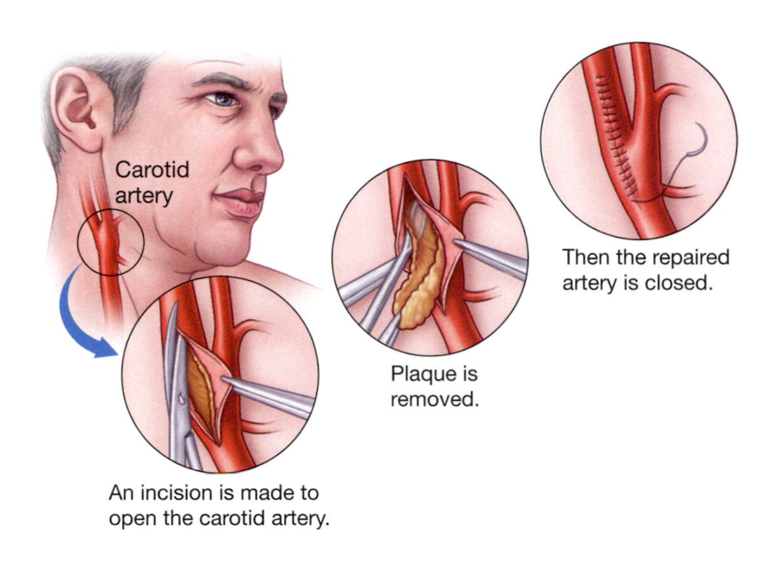

5 *Why does the professor have a slow pulse of 52?*

The professor revealed that he experiences "lecture anxiety." He self-medicates this problem with a beta blocker (Inderal (propranolol), which he purchases over the Internet) every morning before class. This morning, though, he forgot to take the Inderal. He did take it later after his symptoms resolved because he "felt horribly anxious after my mind cleared." The beta blocker reduced his perceived anxiety during class because it blocks the ability of epinephrine to stimulate the B_2 receptors on the heart. As a result, his pulse was reduced and he had less of a sense of "fright or flight." The beta blocker also likely reduced his blood pressure for short periods of time. Beta blockers, however, can impair memory in some people. Also, in people with atherosclerosis, the *immediate* risk of myocardial infarction and/or stroke actually *increases* if the individual skips a daily dose.

The professor might be prescribed a beta blocker as part of his treatment regimen. He will be warned against purchasing prescription medications over the Internet without a prescriber's authorization. Many Internet sites that sell medications without proof of prescription supply medications that are expired or are manufactured without quality control measures. These medications may not contain the stated dosage of the medication.

Finally, the professor will get a referral for behavioral modification to reduce anxiety during public speaking.

References

Johston, S.C., Rothwell, R.M., Nguyen-Huynh, M.N., Giles, M.F., Elkins, J.S., Bernstein A.L., & Sidney, S. (2007). Validation and refinement of scores to predict very early stroke risk after transient ischaemic attack. *Lancet 369*, 283–292.

Transient Ischemic Attack Information Page. National Institute of Neurological Disorders and Stroke
http://www.ninds.nih.gov/disorders/tia/tia.htm

Transient Ischemic Attack. Medline Plus
http://www.nlm.nih.gov/medlineplus/transientischemicattack.html

Panic Free / Fearless Public Speaking Help
http://www.panicfreepublicspeaking.com.au/freetips.html

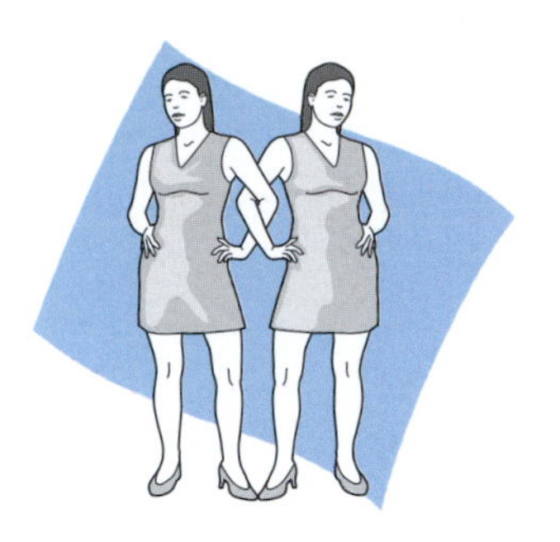

5

Sisters for Life

Mary and Susan Madison are 30-year-old twin sisters who were adopted at age 2. They had a double wedding, and their families live in the same duplex. They even go to medical appointments together. At today's OB/GYN appointment, Susan's blood pressure is normal, 115/70. Her urine is normal. Susan's only complaint is that she has a chronic, intermittent headache in her right temple, which she thinks is a migraine. Nothing much helps the headache, but if she relaxes, the pain usually goes away on its own.

Mary's blood pressure is 165/96. Blood (hematuria) and protein (albuminuria) are found in her urine. Mary reports that she has gained about 15 pounds in the past year and has developed lower back pain and stomach pain that is relieved by acetaminophen (the medicine in Tylenol). The OB/GYN notes that Mary appears to have less muscle mass in her arms and legs and less subcutaneous fat than Susan does. Mary is prescribed an antibiotic for a suspected urinary tract infection, and her urine sample is cultured for confirmation. (Later that week the culture report does come back indicating that Mary had a urinary tract infection.)

Mary's OB/GYN uses ultrasound to examine her back and abdomen. Both of Mary's kidneys are about four times larger than normal and appear to contain numerous fluid-filled cysts. Without further examining Susan, the OB/GYN refers *both* sisters to a nephrologist for MRIs of their kidneys.

Ultrasound Exam of Abdomen

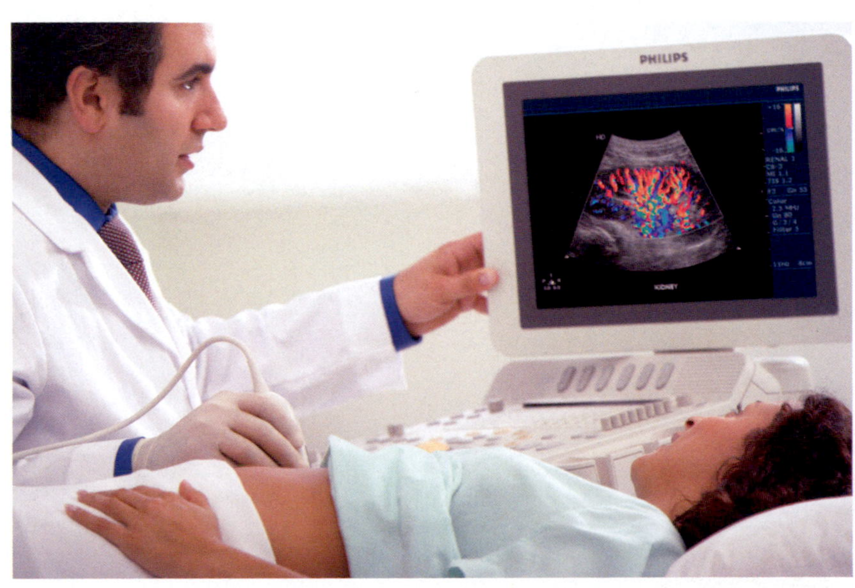

Photo courtesy of Philips Healthcare

Below are the MRI images of Mary and Susan's kidneys.

Mary's MRI

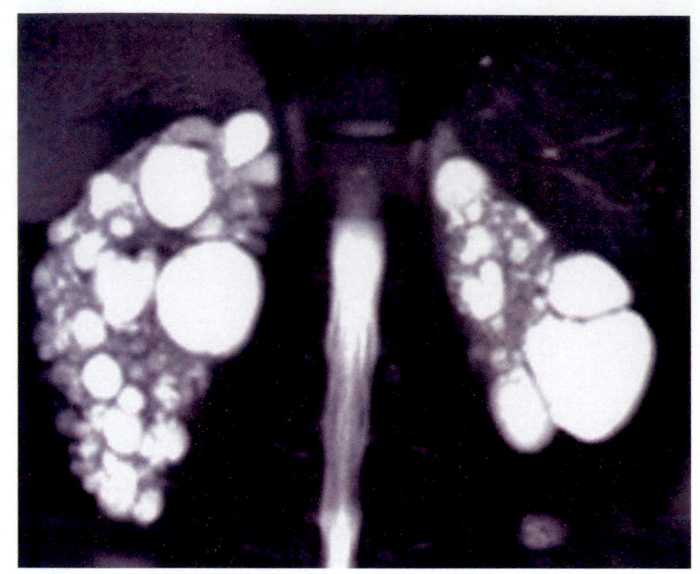

Carol & James Herscot Center for Tuberous Sclerosis Complex and www.massgeneral.org./livingwithtsc

Susan's MRI

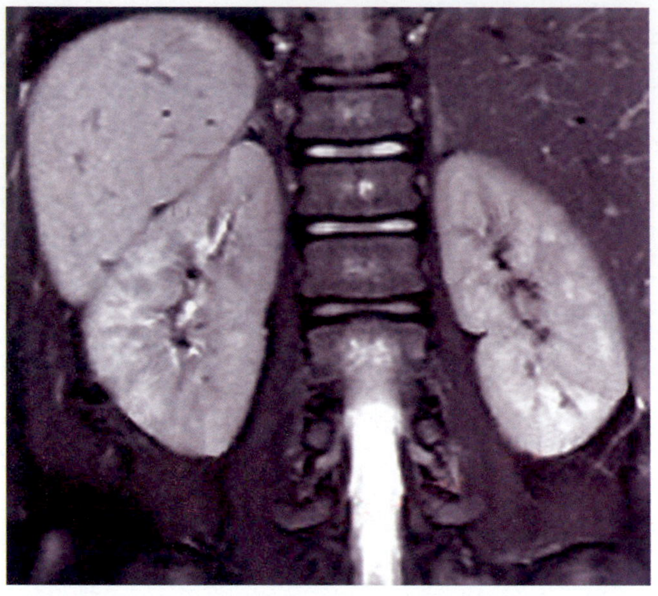

Courtesy of Kyong Tae Bae, MD, PhD, MR radiologist whose research on imaging PKD has been supported by the NIDDK

The nephrologist sends both sisters to have an MRI of the brain and an echocardiogram of the heart. An MRI (magnetic resonance image) is a diagnostic imaging technique whereby the patient is put in a strong, uniform magnetic field. Protons absorb energy from the magnetic field and then emit radio waves as their excitation decays. These radiofrequency signals are converted into three-dimensional images. MRIs do not expose patients to ionizing radiation. An echocardiogram is a series of images of the heart obtained from the reflection or transmission of ultrasonic waves through cardiac tissue.

Mary's MRI of the Brain

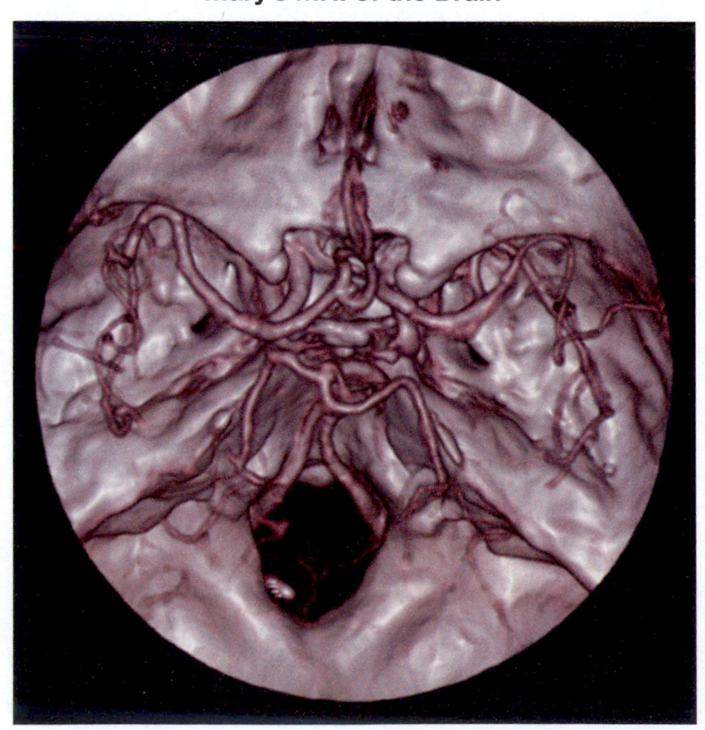

Susan's MRI of the Brain

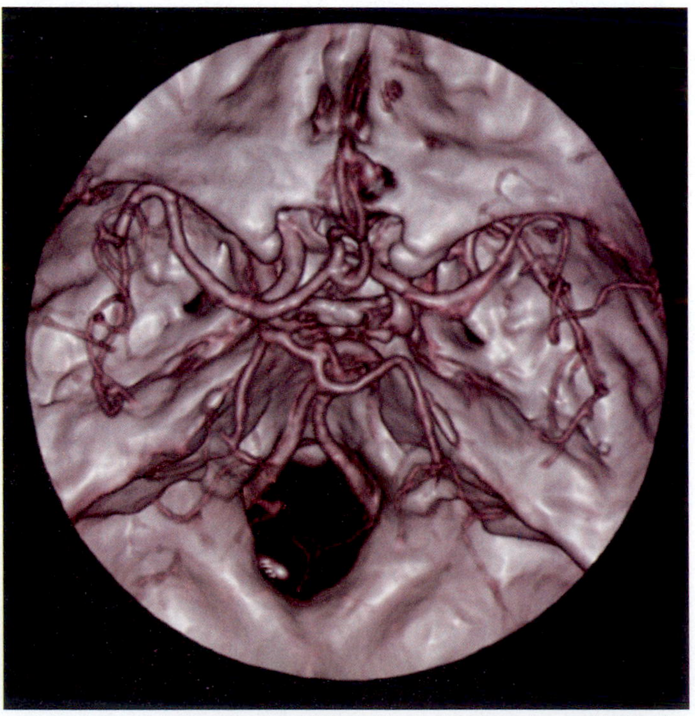

Images courtesy of Hitachi Medical Systems, Europe Holdiing AG

Mary's brain MRI is normal, but Susan's is not. Susan undergoes a cerebral angiogram. A cerebral angiogram involves inserting a catheter into the femoral artery and moving it to the carotid artery. A radio opaque contrast material then is injected. The contrast material blocks the passage of X-rays and allows visualization of blood vessels in the brain on a fluoroscope (an X-ray machine that projects images on a television monitor). Susan's scan is below.

Normal Cerebral Angiogram

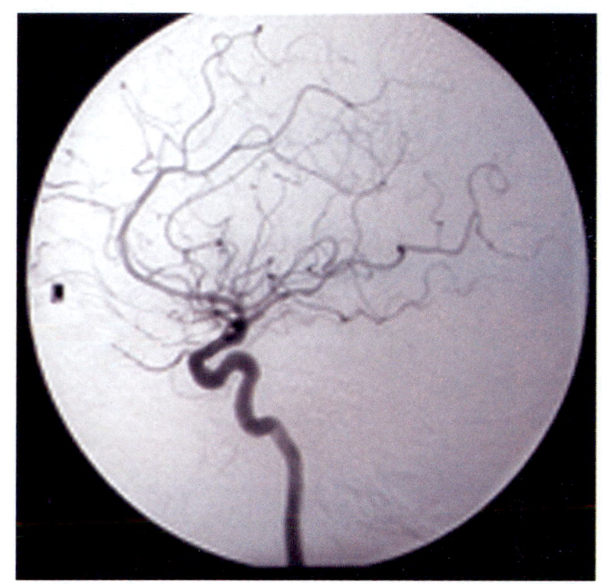

Original image courtesy of Ched Nwagwu, MD

Susan's Cerebral Angiogram

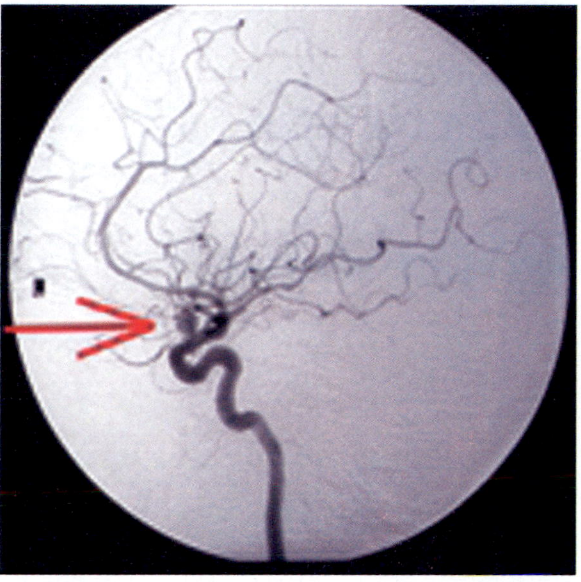

The arrow points to an out-pouching on the cerebral artery.

Image courtesy of Ched Nwagwu, MD

Questions

Name ________________________________ Section __________

1 What problem does Mary have, and how common is it?

2 Why did the OB/GYN send *both* sisters for an MRI of their kidneys?

3 What did the kidney MRIs suggest about Mary? About Susan?

4 Why did Mary gain 15 pounds over the past year?

5 Why does Mary have high blood pressure and a urinary tract infection?

6 Why did the nephrologist send both sisters for a cerebral angiogram and an echocardiogram?

7 Would Susan be a good candidate to give one of her kidneys to Mary?

Answers to Questions

1 *What problem does Mary have, and how common is it?*

Mary has adult polycystic kidney disease (PKD). It is one of the most common inherited disorders and the third most common cause of end-stage renal disease (after hypertension and diabetes).

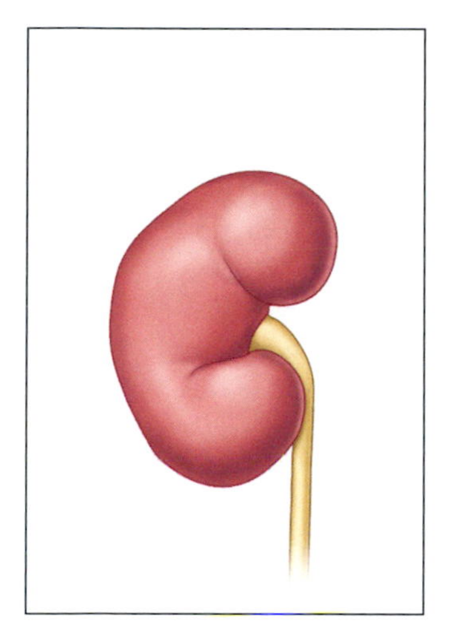

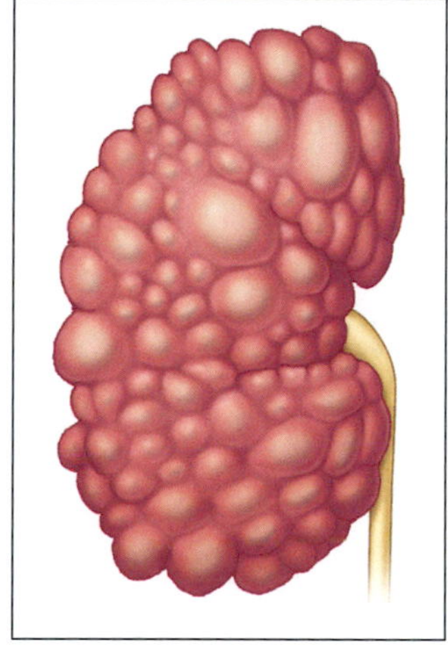

2 *Why did the OB/GYN send both sisters for an MRI of their kidneys?*

PKD diagnosed in a 30-year-old usually is an autosomal dominant disorder that would affect both sisters if they are identical twins. About 25% of cases are new mutations, however. It is unknown if one or both parents had the disease. If both parents had PKD, both sisters would inherit it even if they are fraternal twins. If one parent had this disease, Susan has a 50% chance of inheriting the disease.

3 *What did the kidney MRIs suggest about Mary? About Susan?*

The MRI of Mary's kidneys confirms PKD. Her kidneys are filled with cysts and are more than four times the normal size. MRIs now are used to detect renal cysts in relatives of patients with PKD and to monitor the progression of cysts. Susan does not yet have any cysts. The loss of renal function in patients with PKD has great variability, and only about 50% of individuals with PKD will develop end-stage renal disease requiring renal dialysis or transplant. PKD later is confirmed in Susan with a direct DNA sequence of the PKD1 and PKD2 genes. Susan has a mutation in the PKD1 gene. PKD is a more aggressive disease with PKD1 mutations compared to PKD2 mutations.

Location of PKD1 Gene

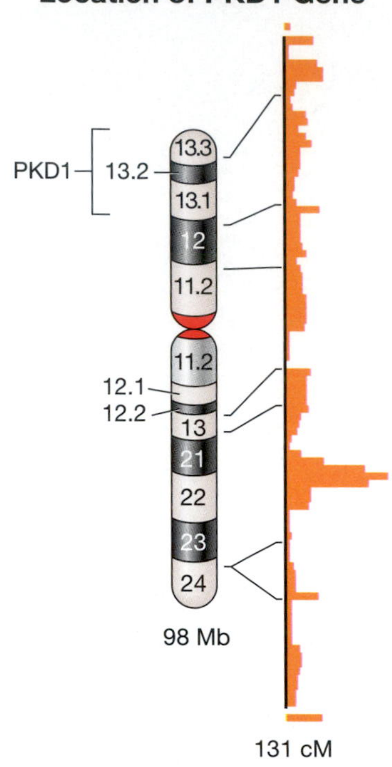

4 *Why did Mary gain 15 pounds over the past year?*

A cyst-filled kidney can weigh 20 pounds or more. Mary's depleted muscle mass and low subcutaneous fat suggest that she actually may have depleted lean body mass at the same time as her kidneys increased greatly in size and weight.

5 *Why does Mary have high blood pressure and a urinary tract infection?*

Often, individuals with PKD develop hypertension before the cysts appear on an ultrasound. Eventually the renal cysts, which begin as out-pouching of a single nephron, detach and enlarge. Cysts can press on the afferent blood vessels that bring blood into the kidney. The decreased renal blood flow triggers renin release, which in turn activates Angiotensin II, which constricts blood vessels and causes hypertension.

Intermittent hematuria affects half of those with PKD. The patient may notice urine that is pink, red, or brown, or red cells may be detected by microscopic examination of a urine sample. The hematuria may be due to irritation from the passage of kidney stones and/or a urinary tract infection. People with PKD have twice the incidence of kidney stones as the general population because cysts block the tubules, allowing urinary stasis and stone formation. Urinary tract infections also are more common in people with PKD. The infections may originate from the bladder (as in most UTIs) or from the cysts themselves. UTIs that originate from the bladder must be treated quickly to prevent the cysts from becoming infected. Certain antibiotics that can penetrate renal cysts are reserved for treating UTIs in people with PKD.

6 *Why did the nephrologist send both sisters for a cerebral angiogram and an echocardiogram?*

About 25% of individuals with PKD have mitral valve prolapse (MVP), compared to about 2%–3% of the general population. Mitral valve prolapse is generally a benign condition, but it can cause mitral regurgitation, skipped heartbeats, chest pain, migraine headache, and anxiety. For 50 years, people with MVP have been told to take a prophylactic antibiotic ("pre-med") before dental procedures and surgery to prevent bacterial endocarditis. The American Heart Association no longer recommends this practice. Mary does not have MVP, but Susan does. This could explain her migraine headaches as migraines are more common in people with MVP.

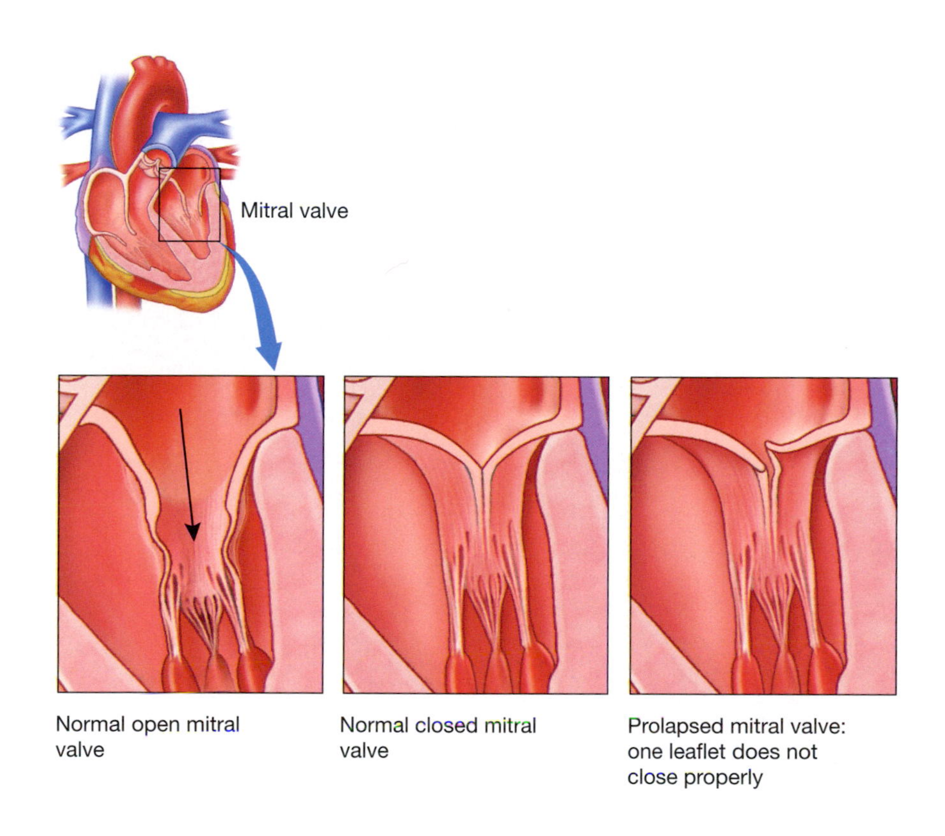

Normal open mitral valve

Normal closed mitral valve

Prolapsed mitral valve: one leaflet does not close properly

About 5%–10% of individuals with PKD have intracranial aneurysms. Susan's brain MRI indicates an aneurysm, and her brain angiogram confirms it. An aneurysm is an out-pouching in a blood vessel that can rupture and cause cerebral hemorrhage. Susan undergoes a procedure called endovascular embolization. Angiography is used to guide a catheter threaded into the femoral artery in the groin up to the aneurysm. Then small spirals of platinum wire are released into the aneurysm and fill it up, causing the blood to clot on the coils and destroy the aneurysm ("coil embolization therapy").

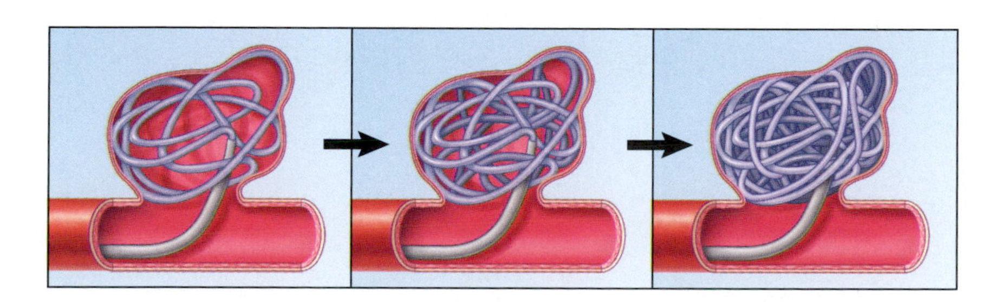

Below is Susan's repeat angiogram on the right. Her previous angiogram is on the left for comparison.

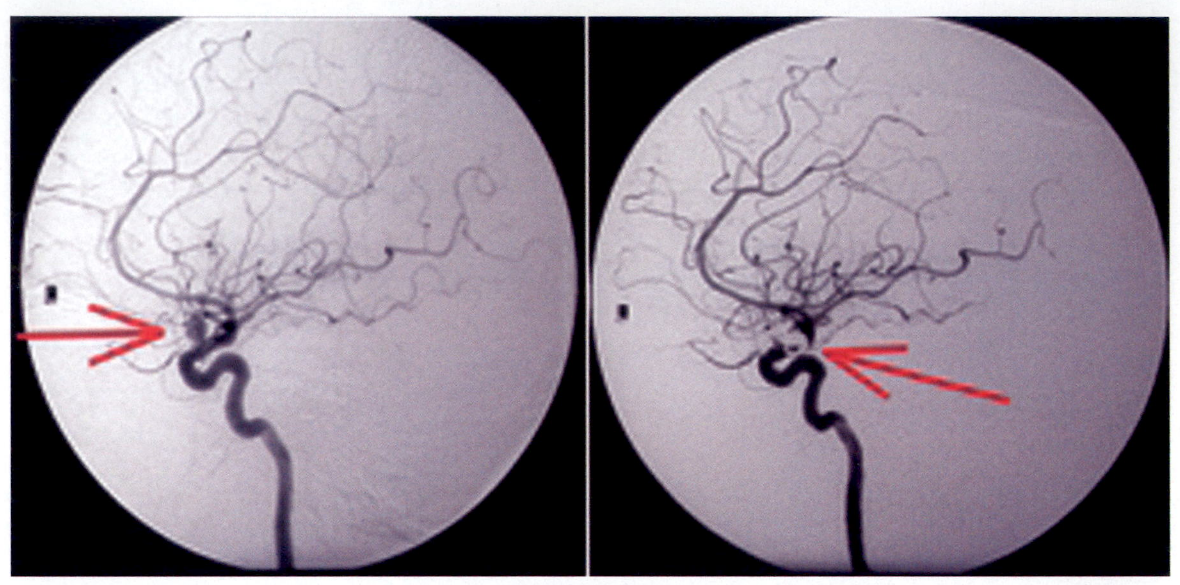

The arrow on the left points to the cerebral aneurysm. The arrow on the right shows the aneurysm after it has been coiled and destroyed. Image courtesy of Ched Nawagwu, MD.

7 *Would Susan be a good candidate to give one of her kidneys to Mary?*

At age 30, Susan does not yet have cysts in her kidneys, but she does have PDK confirmed with DNA sequencing of the PDK genes, and she will develop cysts as she ages. Mary's renal function declines slowly over the next 5 years. At age 35 she goes on renal dialysis and Susan continues to have yearly MRI scans of her kidneys. At age 40, Susan's kidneys are still free of cysts, and she has normal blood pressure. She remains the only good match for a kidney transplant. After 10 years of trying to convince Mary's doctors to let her donate a kidney, they do the procedure.

Both of Mary's kidneys are removed because the cysts are large and cause chronic pain. On the following page is a photograph of Mary's kidneys and the kidney Susan donates. Mary's kidneys are on the left. Susan's kidney is on the right.

The surgery was successful and Mary and Susan remain "sisters for life." They still live next door to each other.

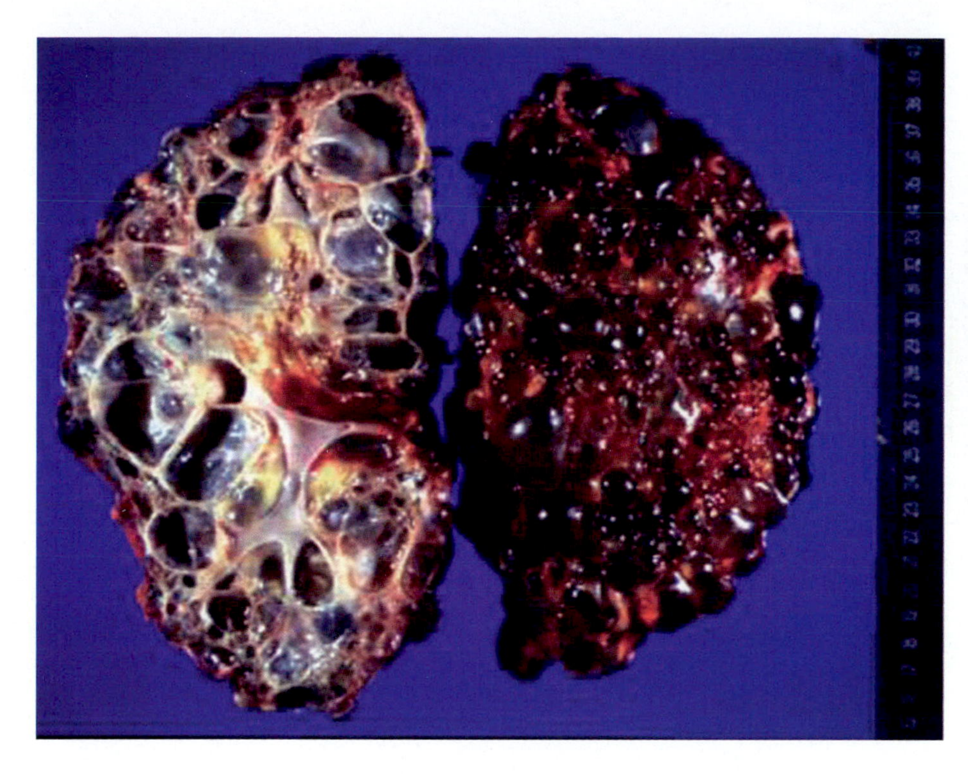

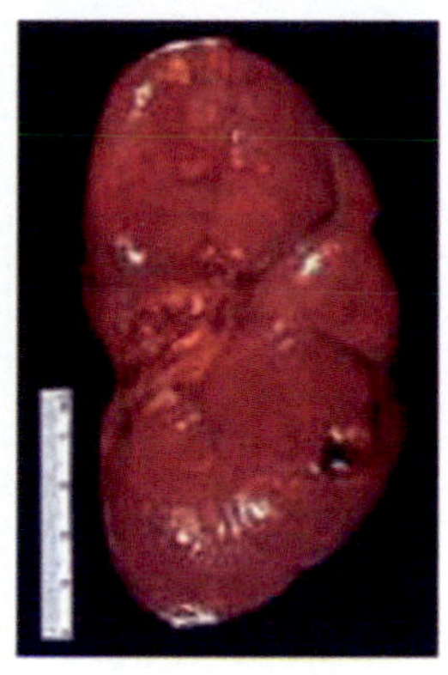

Photos courtesy of the PKD Foundation.

References

PDK Foundation
http://www.pkdcure.org

National Kidney and Urologic Diseases Information Clearinghouse (NKUDIC)
http://kidney.niddk.nih.gov/kudiseases/pubs/polycystic/

National Institute of Neurological Disorders and Stroke: Cerebral Aneurysm Fact Sheet
http://www.ninds.nih.gov/disorders/cerebral_aneurysm/detail_cerebral_aneurysm.htm#_ftnref1

The American Heart Association: New guidelines regarding antibiotics to prevent infective endocarditis. May, 2007.
http://www.americanheart.org/presenter.jhtml?identifier=3047051

6

The Hoarse Horsewoman

Brittany Shores is an accomplished equestrian who volunteers 20 hours per week teaching children with special needs how to ride and care for horses. She is in the Ear, Nose, and Throat Clinic today. She says that "by Wednesday, I have no voice. This has been going on for about four months now. Last Monday I woke up and literally couldn't speak. I have a chronic cough, too, but I've had that for years. It's from breathing hay. I usually wear a mask when I muck out the barn."

A physical exam and medical history reveals the following:

Age: 51
Height/weight: 5'4" 140 lbs (has lost 15 lbs in the past year)
Blood pressure: 115/70
Pulse: 64
Former cigarette smoker (1 pack/day from ages 18–28)
Chronic cough produces scant sputum with blood streaks
Adopted, no family history available

A chest X-ray is ordered and is shown below:

Normal X-ray

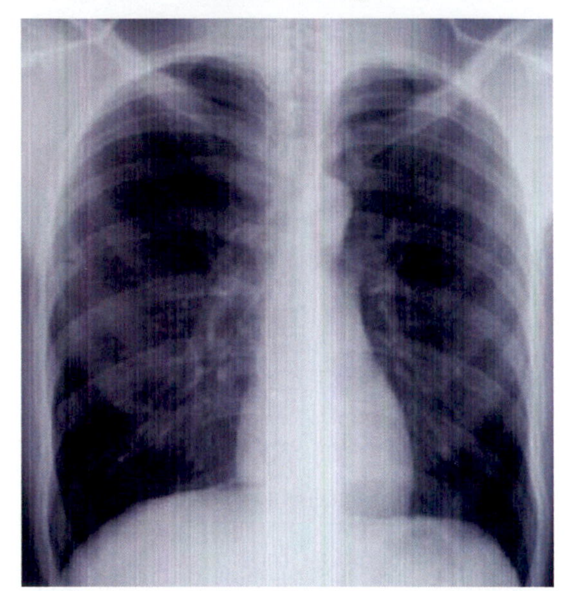

Brittany's X-ray

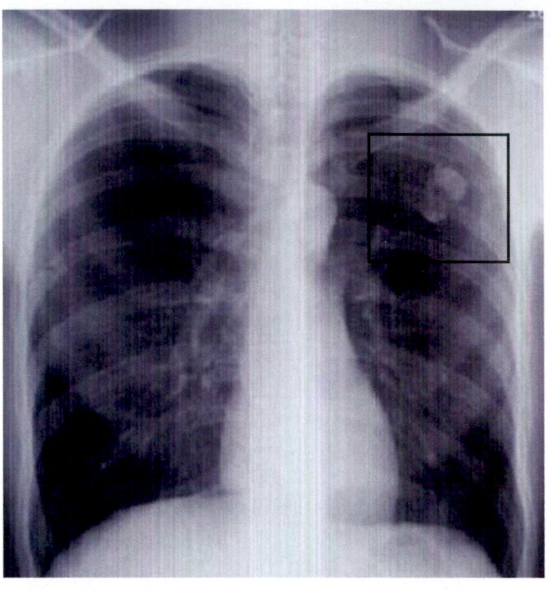

Reproduced with permission from: Stark, P. Evaluation of diffuse lung disease by plain chest radiography. In: UpToDate, Basow, DS (Ed.), UpToDate, Waltham, MA 2009. Copyright 2009 UpToDate, Inc. For more information visit www.uptodate.com.

Following the X-ray, a spiral CT scan is ordered. A spiral CT scan is different from a regular CT scan in that the patient is moved rapidly through the spiral CT scanner. Spiral CT scans produce images with higher definition of blood vessels and internal tissues within the chest cavity. Images from the spiral CT scan are shown below. (Note A is aorta, S is spine.)

Normal Spiral CT Scan

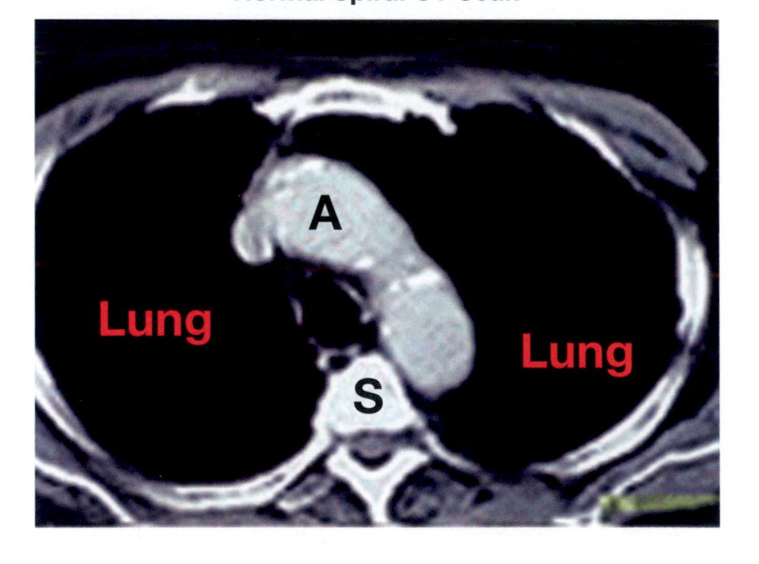

Brittany's Spiral CT Scan

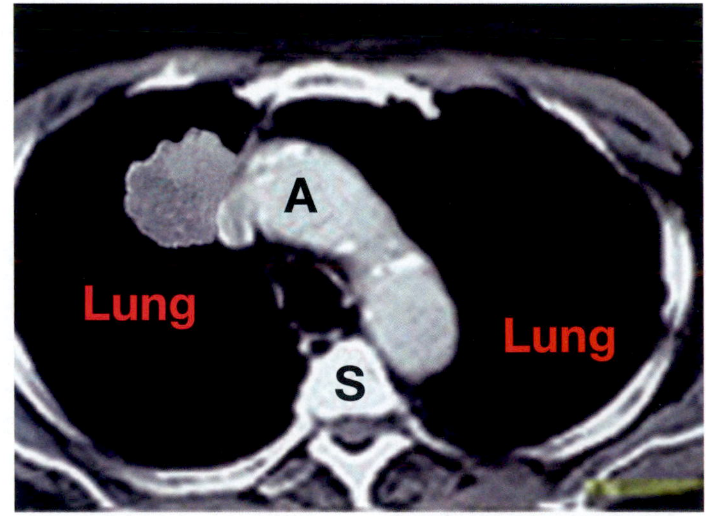

Images courtesy of The National Library of Medicine

Questions

Name ______________________________ Section __________

1 What underlying problem does Brittany Shores appear to have?

2 How common is this problem in women?

3 What are the possible causes of this problem?

4 Why does this problem present as laryngitis? (Consider the anatomy of the laryngeal nerve in your answer.)

5 What are the screening tests for early diagnosis of this problem?

6 What are the clinical signs of this problem?

7 What are the treatment options for this problem?

Answers to Questions

1 *What underlying problem does Brittany Shores appear to have?*

She has a mass in her left lung consistent with lung cancer. There are four major types of lung cancer: small cell, adenoma, large cell, and squamous. A biopsy reveals that Brittany Shores has small cell cancer.

2 *How common is this problem in women?*

Lung cancer now kills more than 67,000 women per year—more than breast, ovarian, and uterine cancer combined.

3 *What are the possible causes of this problem?*

Cigarette smoking (including passive exposure to second-hand cigarette smoke) causes about 85%–90% of lung cancers. Only 10%–15% of lung cancers occurs in nonsmokers. Other risk factors for lung cancer include:

- lung diseases such as tuberculosis or chronic obstructive pulmonary diseases such as chronic bronchitis or emphysema
- exposure to asbestos
- exposure to radon (Radon is a naturally occurring gas that distributes into the air through the basement. It also moves into water and can be inhaled during showering.)
- exposure to airborne contaminants (such as diesel fuel, air pollution)
- a diet with a low intake of fruits, vegetables and grains and a high intake of fat and cholesterol
- certain genes that promote cancer growth in response to estrogen

Women are more susceptible than men are to lung cancer. Women develop lung cancer earlier than men, and with less smoke exposure. They also are more likely to develop small cell lung cancer and adenocarcinoma, both of which metastasize before symptoms develop, and they have the poorest prognosis. Men are more likely to develop squamous cell carcinoma in the bronchi, which can be detected earlier than other types of lung cancer.

4 *Why does this problem present as laryngitis?*

Lung cancer in the left lung apex or mediastinum (primary tumor or lymph node metastasis) can invade the laryngeal nerve and cause nerve palsy (paralysis of the recurrent laryngeal nerve). As a result, the patient will experience hoarseness or laryngitis.

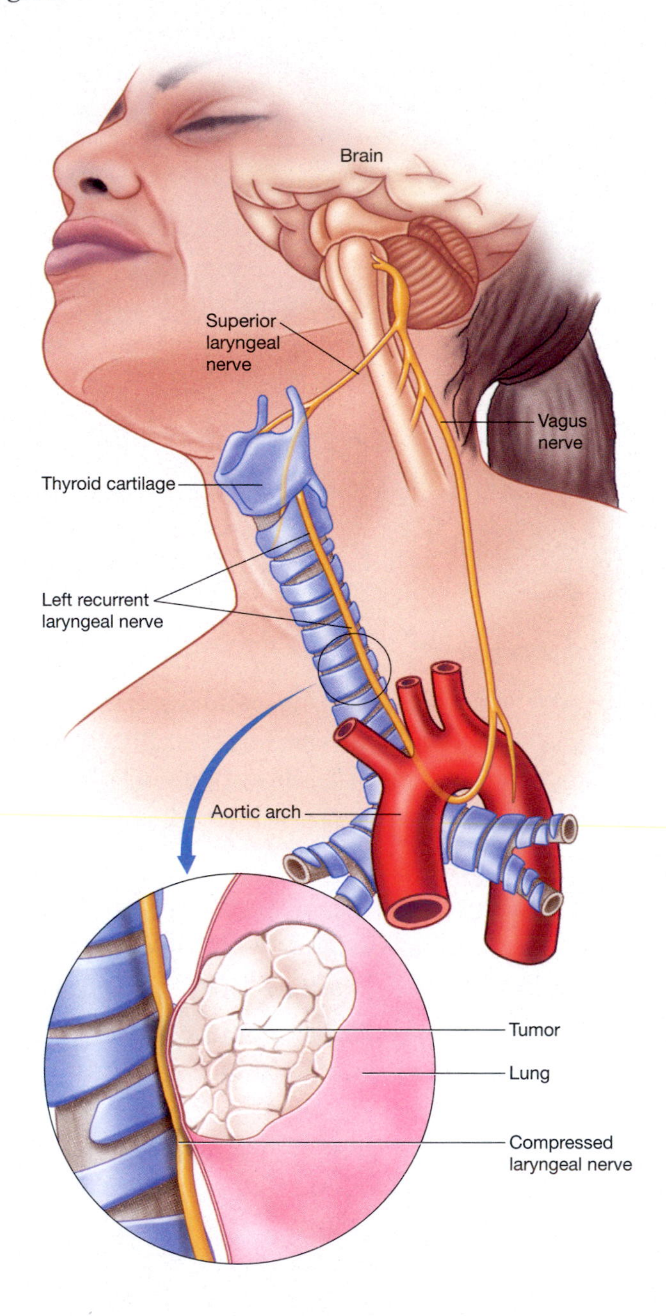

5 *What are the screening tests for early diagnosis of this problem?*

There are no early screening tests for lung cancer. Early lung cancer is not visible on chest X-ray. Sputum samples are normal in early lung cancer, and so are also non-diagnostic.

Computed tomography (CT) scans can diagnose lung cancer before symptoms develop, but screening smokers with CT scans has not yet been proven to reduce mortality from lung cancer. Currently, routine CT scans are not recommended for the general population. Many clinicians, however, now recommend periodic CT scans for their patients who smoke.

6 *What are the clinical signs of this problem?*

Symptoms of lung cancer include:

- a persistent cough that worsens over time
- sputum that contains blood
- persistent chest pain
- hoarseness
- shortness of breath or wheezing
- difficulty swallowing
- recurrent upper respiratory infections
- swelling of the neck or face
- fatigue, loss of appetite, and weight loss

7 *What are the treatment options for this problem?*

Small cell lung cancer (also called oat cell cancer) is a systemic lung cancer in about 70% of cases by the time it is diagnosed. Surgery is rarely an option. The 5-year survival rate is less than 20%. Treatment consists of chemotherapy and palliative radiation to shrink the tumors and lymph nodes so they are less likely to impinge on nerves and organs.

After getting the diagnosis, Ms. Shores reveals, "We had our water tested for radon when we bought our horse farm four years ago. It was really high, so we had an aeration system installed. I wonder now if the house I grew up in (a mile down the road) had radon in the water. Also—I guess I should have told you this—but I do use marijuana, about six joints a week or so. I've used marijuana since I was 20." Recent studies (see references) suggest that smoking marijuana does not increase the risk of lung cancer. However, the possible interaction of marijuana, diet, and exposure to radon has not yet been evaluated. Although Ms. Shores gave up smoking 23 years ago, both cigarette smoking and radon likely played a role in her developing lung cancer.

References

Spiral CT Scans for Lung Cancer Screening: Fact Sheet, National Cancer Institute
http://www.cancer.gov/cancertopics/factsheet/lung-spiral-CTscan

Lung Cancer. Harvard Health
http://www.health.harvard.edu/articles/lung_cancer.htm

Maghfoor, I. (2008). Lung Cancer, Oat Cell (Small Cell). Updated Oct. 15, 2008
http://eMedicine.medscape.com/article/280104-overview

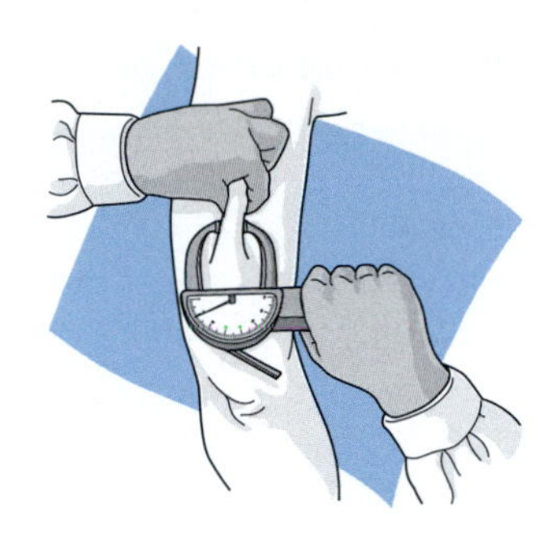

7

The Perfect Diet

Sharon Rock, age 21, is hospitalized with severe abdominal pain. The pain is a "10" on a 0–10 pain scale but is intermittent. She vomits after each pain wave. Her abdomen is distended and measures 38 inches. A clinical dietitian measures her mid-arm circumference and triceps skinfold thickness. She is given a pregnancy test. A flat plate X-ray of her abdomen shows air under the diaphragm. Another X-ray shows a "density" (an area that blocks the X-rays) in her duodenum. She has tachypnea (rapid breathing) and dsypnea (difficulty breathing). When a stethoscope is placed over her lower abdomen, there are no bowel sounds.

Here are some of Sharon's test results:

Height: 5'5"
Weight: 100 lbs
Pregnancy test: Negative

		Normal Values
Mid-arm circumference	18 cm	28.5 cm
Triceps skinfold	8 mm	16.5 mm
Potassium	5.2 mEq/L	3.5–5.0 mEq/L
Albumin	3.0 g/100ml	4.0–5.5 g/100ml

X-ray Showing Air Under the Diaphragm

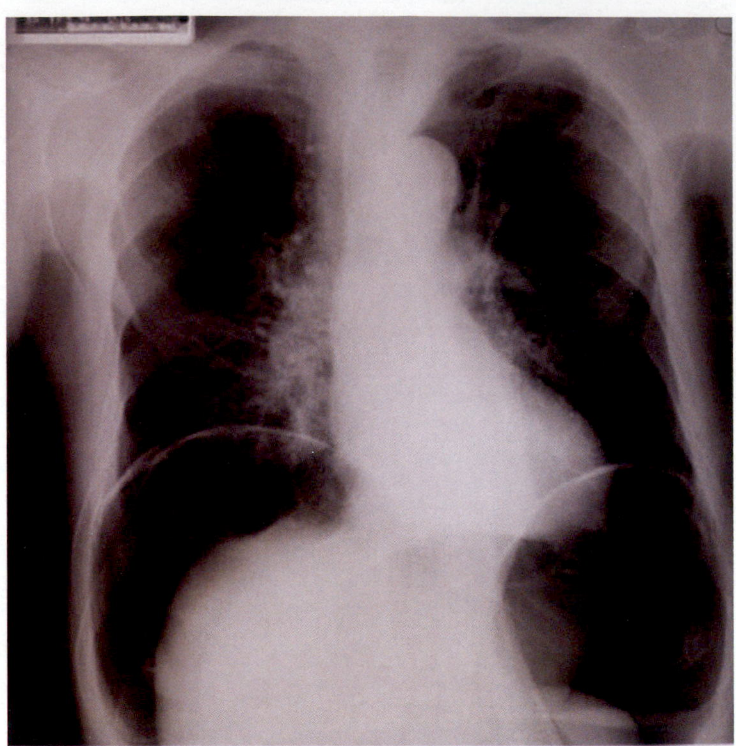

Image courtesy of McGill Molson Medical Informatics Project

X-ray Showing Density in Small Intestine

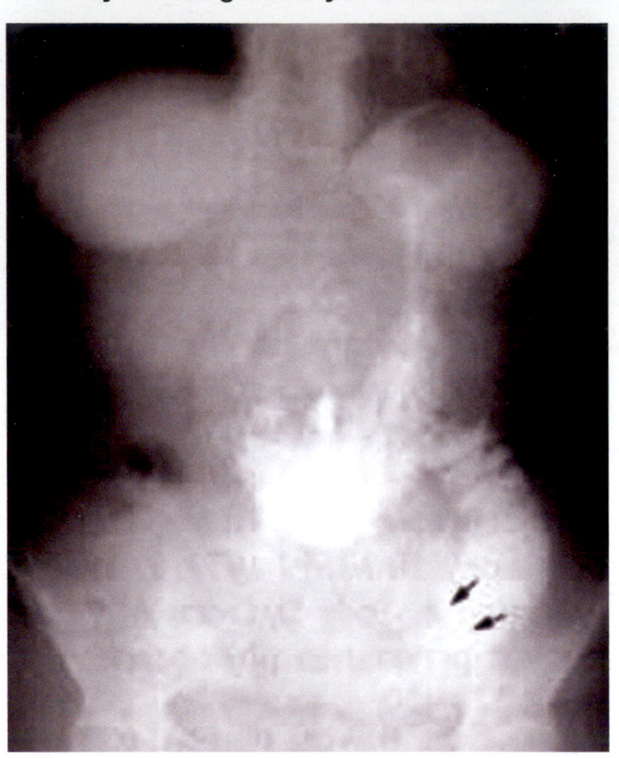

Used with permission from the Medical University of South Carolina's Digestive Disease Center

Questions

Name ______________________ Section __________

1 What do the mid-arm circumference and triceps skinfold results suggest?

2 What do the X-rays suggest?

3 Why is her abdomen distended? (Give two reasons!)

4 Why is her potassium high?

5 Why does she have tachypnea and dyspnea?

6 Sharon is brought in for emergency surgery. What do you expect the surgeon will find?

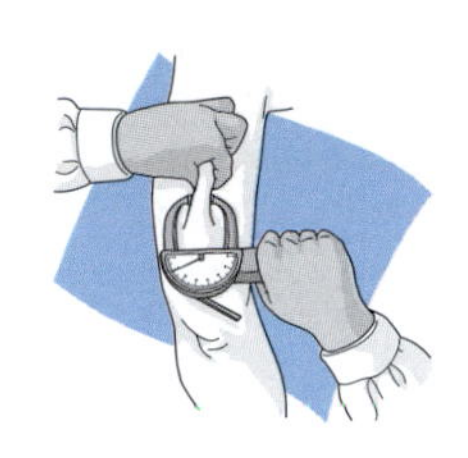

Answers to Questions

1 *What do the mid-arm circumference and triceps skinfold results suggest?*

Sharon's skeletal muscle mass and subcutaneous fat are both depleted. Her weight-to-height ratio is low. She may have anorexia nervosa and/or bulimia. The following formula is used to calculate the mid-arm muscle circumference (MAMC) from the mid-arm circumference (MAC) and triceps skinfold (TSF):

$$\text{MAMC} = \text{MAC (cm)} - (3.14 \times \text{TSF (cm)}) \quad \text{(Normal: 23.2 cm)}$$

$$\text{Sharon's MAMC} = 18\text{ cm} - (3.14 \times 0.8\text{ cm}) = 15.49\text{ cm}$$

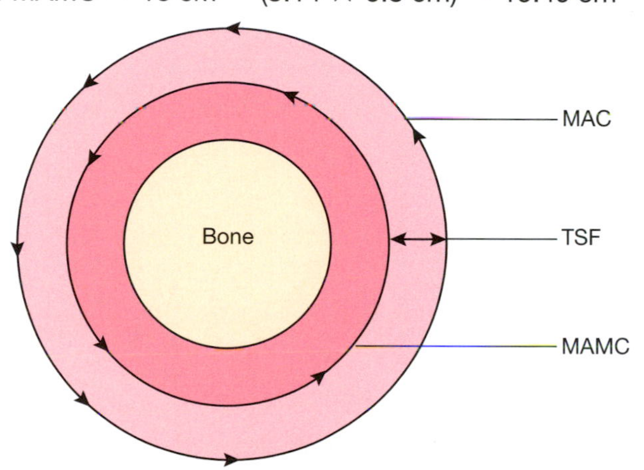

The mid-arm muscle circumference indicates how much somatic (muscle) protein an individual has. Sharon's MAMC of 15.49 cm indicates significant depletion of somatic protein. In addition, her depleted albumin suggests decreased visceral (internal) protein. Taken together, these lab values indicate that Sharon is at greater risk for infection and poor wound healing following surgery. This is because if there is an insufficient energy (calorie) intake, the body will use somatic and visceral proteins to meet energy needs at the expense of synthesizing antibodies and tissue proteins.

2 *What do the X-rays suggest?*

Sharon has an intestinal obstruction. The air under the diaphragm indicates that she has a perforation. Swallowed air enters the gastrointestinal (GI) tract, exits through the perforation and collects under the diaphragm.

3 *Why is her abdomen distended?*

Her serum albumin is low. Albumin provides colloid osmotic pressure (also known as oncotic pressure) to hold water in the blood. Without it, water moves from the blood to interstitial tissues and causes ascites. The ascites (fluid accumulation) is in the peritoneal cavity, the space between the peritoneal membrane (peritoneum) and the abdominal organs. Also, the obstruction prevents water from being absorbed in the colon. The obstruction plus unabsorbed water distends the intestines.

4 *Why is her potassium high?*

Her potassium is high as a result of water moving to interstitial tissues. This process concentrates electrolytes in the blood. Also, the obstruction is preventing water from reaching the colon, where the greatest percentage of water is normally absorbed. Less absorbed water concentrates electrolytes in the blood as well.

5 *Why does she have tachypnea and dyspnea?*

The air pressing on the diaphragm in turn presses on the base of the lungs and decreases her lung volume. As a result, she has an increased respiratory rate, called tachypnea, and she has difficulty breathing, called dyspnea.

6 *Sharon is brought in for emergency surgery. What do you expect the surgeon will find?*

Before she is put under anesthesia, Sharon says, "I thought I had the perfect diet. I eat four fiber wafers for supper so I'm not hungry at night. I've lost weight and kept it off. I was so pleased."

Taking fiber at night without sufficient fluid can cause an intestinal obstruction. Peristalsis slows during the night, and the fiber can coalesce and result in an undigested mass called a bezoar. Surgical removal and resection of the intestine can lead to complete recovery—if Sharon also attends to her disordered eating behaviors.

Fiber Bezoar

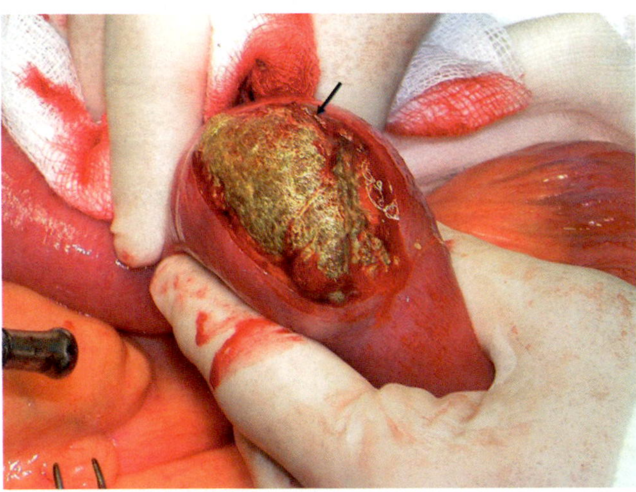

Reprinted with permission from Teng HC, Nawaw O, Ng KL, Yik YI, Phytobezoar: An Unusual Cause of Intestinal Obstruction, Biomed Imaging Interv J 2005;1(1):e4 www.biij.org/2005/1/e4/

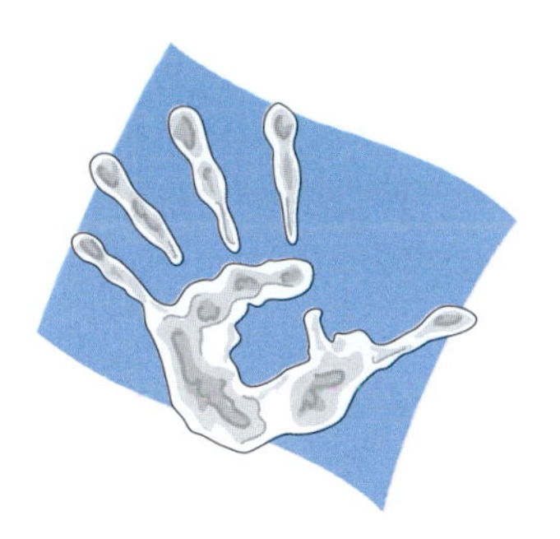

8

A Failed Preschool Art Project

Justin Jones, age 4, is being seen in a pediatric hospital clinic. Mrs. Jones says, "Tuesday was Justin's first day of preschool. The children made handprint cards for their parents. Justin's teacher sent a note home to tell us that Justin's project didn't turn out too well because he doesn't have any fingerprints! We were bringing Justin to the clinic today because he just doesn't seem to be growing. He is so much smaller than the other children in his class. And look at us—we're both tall! He has some kind of skin allergy on his arms, too. But now we're more worried about why he has no fingerprints!"

Justin appears small for his age. (His growth data follow on page 65.) All other physical parameters appear to be within normal limits. His mother reports that Justin "has a good appetite, but he gets stomach pains and then diarrhea about two hours after he eats. The diarrhea is nasty. It's sort of frothy, and it floats!" He has patches of severe itchy skin on his upper arms, and his fingerprint ridges, though present, are faint.

Justin's Upper Arm

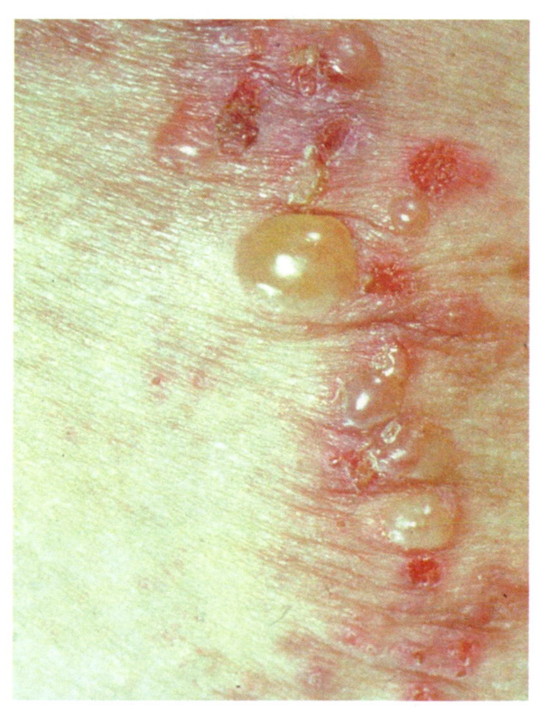

Reprinted from *Dermatology Nursing*, 2004, Volume 16, Number 1, p 31. Reprinted with permission of the publisher, Jannetti Publications, Inc., East Holly Avenue, Box 56, Pitman, NJ 08071-0056; Phone (856) 256-2300; Fax (856) 589-7463. (For a sample of the journal, please contact the publisher or visit www.dermatologynursing.net)

Justin undergoes an endoscopic exam, and a biopsy is taken from his jejunum.

Drawing of Normal Villi

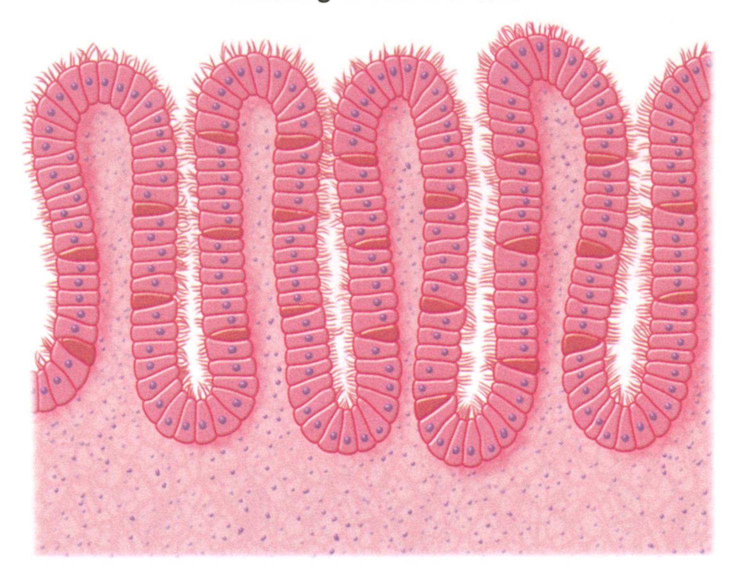

Drawing of Justin's Villi

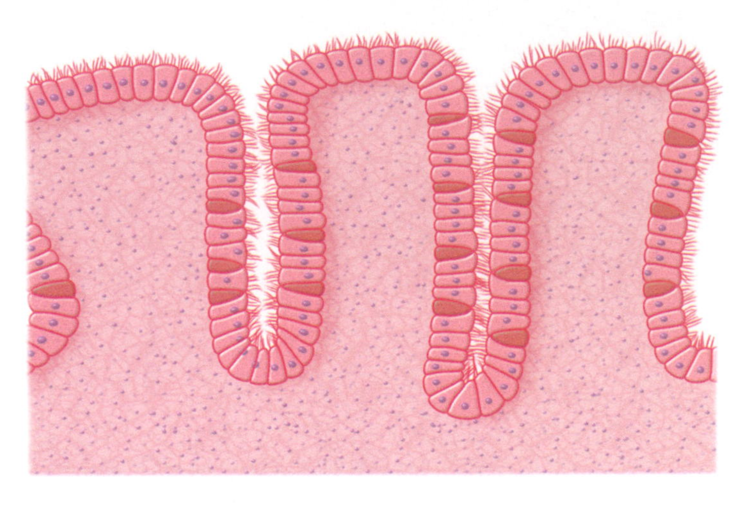

Micrograph of Normal Villi

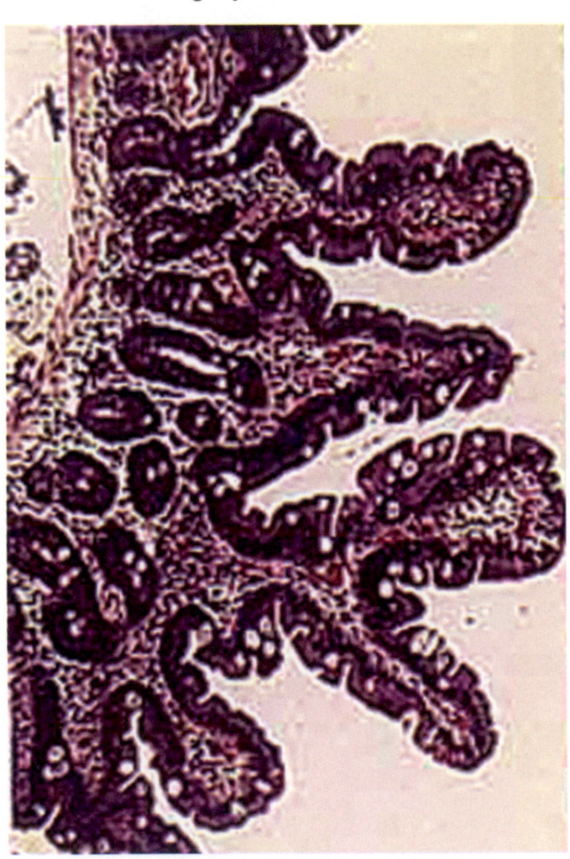

Micrograph of Justin's Villi

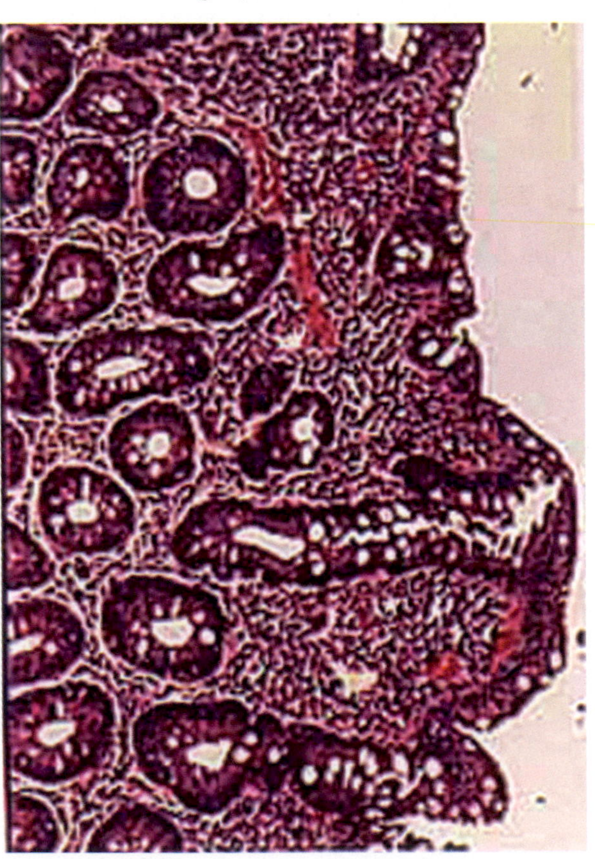

Figure 3 from "Celiac Disease" [Presutt, RJ, Cangemi JR, Cassidy HD, Hill, DA-author(s); American Family Physician 76(2)2170; December 15, 2007]

Questions

Name ______________________________ Section __________

1 What percentile for height and weight is Justin? Plot Justin's growth on the chart below. Enter "Justin Jones" after name. Enter record # as 3120. Enter the following data for:

Date	Age	Weight	Stature
8/1/06	2 yr 2 mo	12 kg	86 cm
9/1/07	3 yr 3 mo	12 kg	90 cm
9/7/08	4 yr 3 mo	14 kg	99 cm

NAME ______________________

Weight-for-stature percentiles: Boys

RECORD# __________

Date	Age	Weight	Stature	Comments

kg: 8, 9, 10, 11, 12, 13, 14, 15, 16, 17, 18, 19, 20, 21, 22, 23, 24, 25, 26, 27, 28, 29, 30, 31, 32, 33, 34

lb: 20, 24, 28, 32, 36, 40, 44, 48, 52, 56, 60, 64, 68, 72, 76

Percentile curves: 97, 90, 85, 75, 50, 25, 10, 3

STATURE

cm: 80, 85, 90, 95, 100, 105, 110, 115, 120

in: 31, 32, 33, 34, 35, 36, 37, 38, 39, 40, 41, 42, 43, 44, 45, 46, 47

Published May 30, 2000 (modified 10/16/00).
SOURCE: Developed by the National Center for Health Statistics in collaboration with the National Center for Chronic Disease Prevention and Health Promotion (2000).
http://www.cdc.gov/growthcharts

2 Describe how Justin's villi differ from normal.

3 What functions of villi are impaired as a result of Justin's condition?

4 What caused the diarrhea? What is the technical term for this type of diarrhea?

5 How does Justin's skin differ from normal?

6 What condition does Justin have, and what causes it?

7 What is the etiology of this condition, and how frequent is it?

8 What other disease is frequently associated with Justin's condition?

9 What diseases will Justin be at increased risk for over the long term?

10 How can this condition be treated?

Answers to Questions

1 *What percentile for height and weight is Justin? Plot Justin's growth on the chart on the following page.*

Enter Justin Jones after name. Enter record # as 3120. Enter the following data for:

Date	Age	Weight	Stature
8/1/06	2 yr 2 mo	12 kg	86 cm
9/1/07	3 yr 3 mo	12 kg	90 cm
9/7/08	4 yr 3 mo	14 kg	99 cm

Justin was just below the 50th percentile for height and weight at age 2 years 2 months, suggesting that he was at an average weight-for-stature for his age. His weight-for-stature fell to the 10th percentile at ages 3 years 3 months and at 4 years 3 months. His fall-off in height and weight suggests a failure to grow.

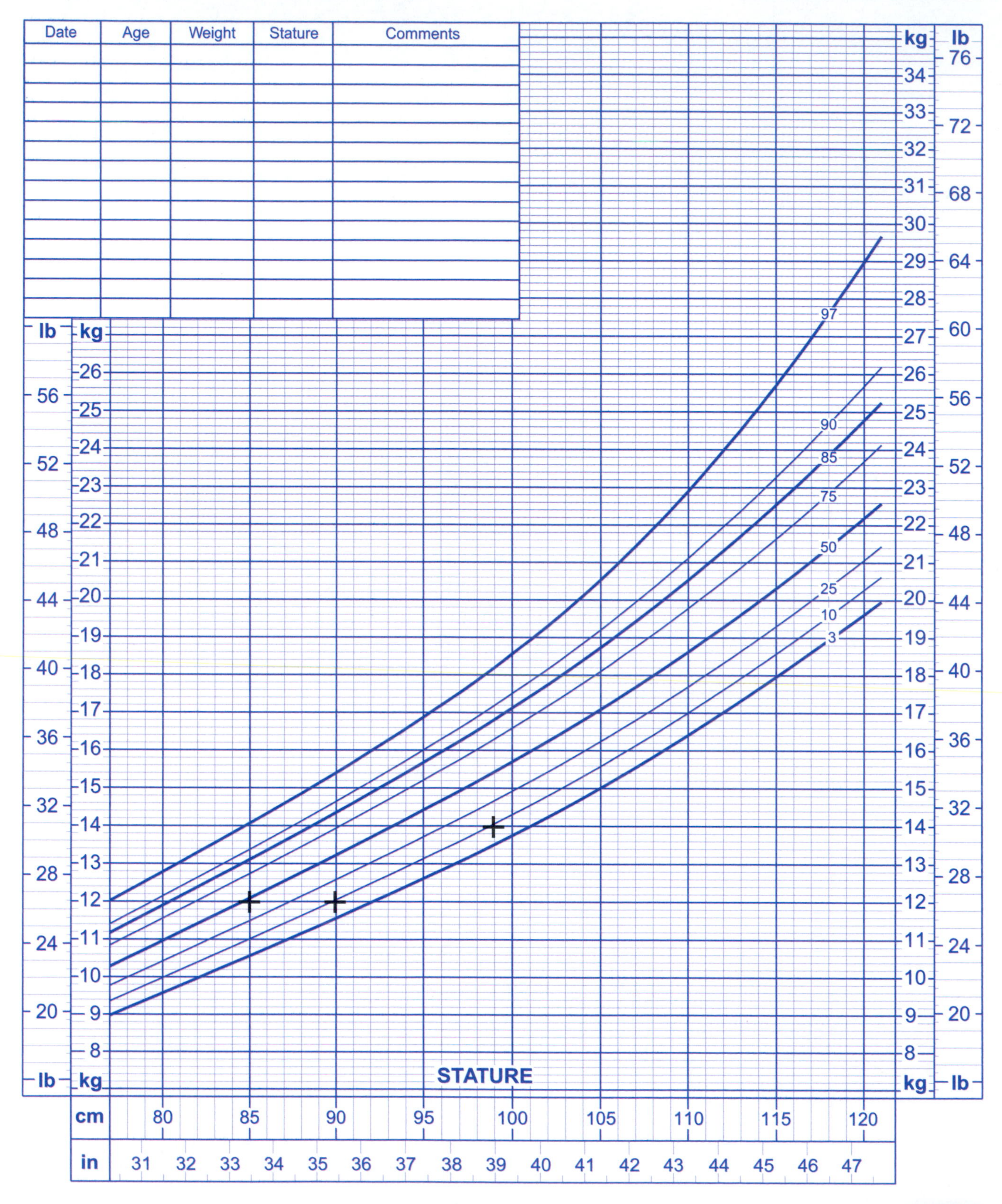

Published May 30, 2000 (modified 10/16/00).
SOURCE: Developed by the National Center for Health Statistics in collaboration with the National Center for Chronic Disease Prevention and Health Promotion (2000).
http://www.cdc.gov/growthcharts

2 *Describe how Justin's villi differ from normal.*

Notice that the villi are shorter and the tops are flatter. Also, the cells appear disorganized.

3 *What functions of villi are impaired as a result of Justin's condition?*

Justin's villi have a smaller surface area. This will reduce his ability to absorb nutrients such as amino acids (from protein), vitamins, fat, and lactose.

4 *What caused the diarrhea? What is the technical term for this type of diarrhea?*

Unabsorbed nutrients caused the diarrhea. The unabsorbed nutrients caused the intestinal contents to have a high osmotic pressure, keeping water in the fecal material. Fecal material described as "frothy" and "floats" typically contains fat. This is called steatorrhea. ("Steatos" is Greek for fat.)

5 *How does Justin's skin differ from normal?*

Justin's skin has inflammatory lesions that are red (erythematous), raised (papular), with puss-filled lesions (pustules), and blisters (vesicles). A skin biopsy shows IgA in the tissues. This confirms dermatitis herpetiformis, which is associated with Justin's primary condition. The IgA particles obstruct small capillaries in the skin and attract neutrophils, which in turn cause an inflammatory response in the skin.

6 *What condition does Justin have, and what causes it?*

Justin has celiac disease, also known as gluten sensitive enteropathy. It is caused by an autoimmune reaction to a protein called gliaden, which is contained in many grains such as wheat, barley, rye, malt, and triticale (but not rice or oats). Symptoms include failure to thrive, with gastrointestinal pain, steatorrhea, atrophy of the fingerprint ridges, and dermatitis herpetiformis. Diagnosis can be confirmed with a biopsy and IgA/IgG antibody tests against gluten (both IgA and IgG antibodies can bind to gluten) protein.

7 *What is the etiology of this condition, and how frequent is it?*

One in 133 people has celiac disease, but the majority of these people remain undiagnosed.

8 *What other disease is frequently associated with Justin's condition?*

Type I diabetes (formerly known as juvenile diabetes or insulin-dependent diabetes) is associated with celiac disease.

9 *What diseases will Justin be at increased risk for over the long term?*

People with poorly controlled celiac disease have an increased risk of a type of cancer called intestinal lymphoma. He will also be at increased risk for failure to grow, anemia (from iron, folate, and B_{12} deficiency), and osteoporosis (from calcium and vitamin D deficiency).

10 *How can this condition be treated?*

Complete avoidance of gluten-containing products will lead to recovery of Justin's villi. His growth likely will catch up within several months, and his skin condition will improve. Gluten is in many foods as a component of the food itself (e.g., pasta, bread) or as an ingredient (e.g., wheat extract in soy sauce). Many gluten-free products such as breakfast cereal, bread, and pasta made from rice are now available. Examples of foods that Justin will have to avoid for the rest of his life include the following.

Foods to Avoid in Celiac Disease

wheat	barley	rye	triticale	malt
bread	pasta	crackers	cold cuts	stuffing
cheese products	soy protein	meat substitutes	Communion wafers	
baked beans	gravies	salad dressings	soups	

References

Celiac Disease Foundation
http://www.celiac.org/

American Celiac Disease Alliance
http://www.americanceliac.org/

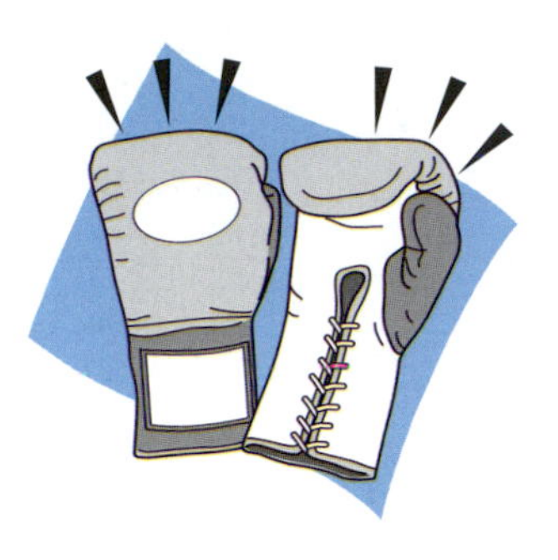

9

Kidney Punch

Gerry M., age 23, is an amateur boxer and lays carpet for a living. He states that he had a kidney stone about three weeks ago and went to the emergency department (ED) because the pain was so bad. The stone passed while he was waiting for treatment. He felt fine after that and went back to his previous routine. Then, two weeks ago, his left elbow was sore, red, and swollen when he woke up in the morning. He went to the ED and was diagnosed with olecranon bursitis and given a prescription for ibuprofen (800 mg four times a day). Today he is back at the ED complaining of severe pain in his lower back (in the area of the costovertebral angle on both sides), and his bursitis has not resolved.

Normal Olecranon Bursa

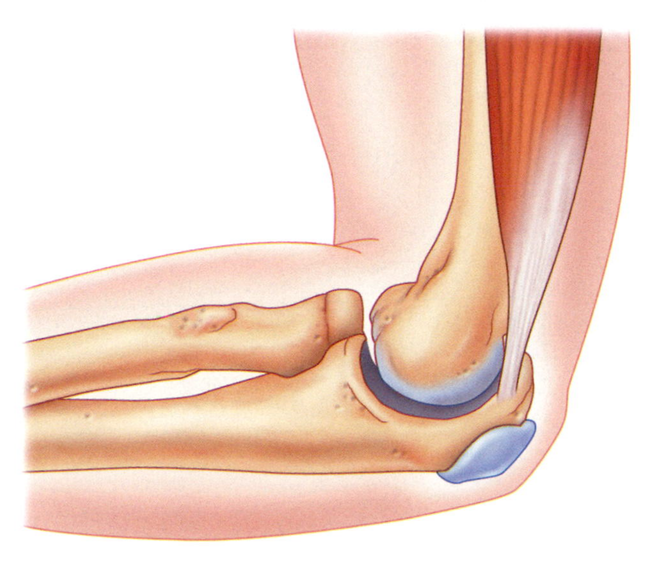

Olecranon Bursitis

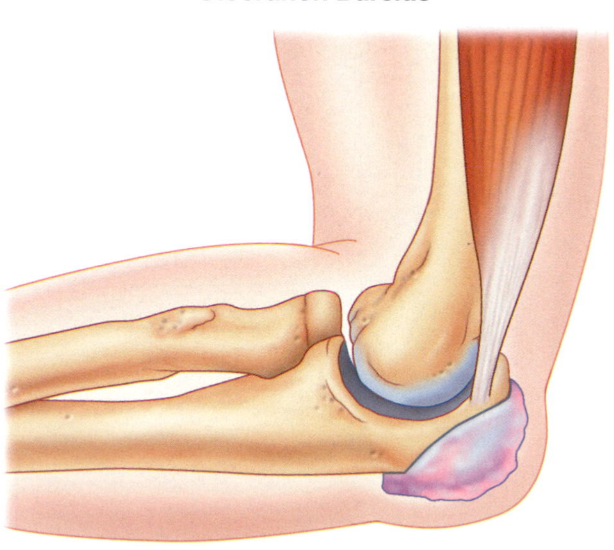

Bursal fluid is aspirated from his left elbow and analyzed.

He is given an axial CT scan of his lower abdomen. In an axial CT scan, a CT scan is taken, and then the table on which the patient is situated is moved a tiny bit and another scan is taken. This process is called "step and shoot." The CT scan machine then uses tomographic reconstruction to generate a 3-D axial image.

Gerry M. goes into acute renal failure and is placed on the kidney/liver transplant list.

Axial CT Scan

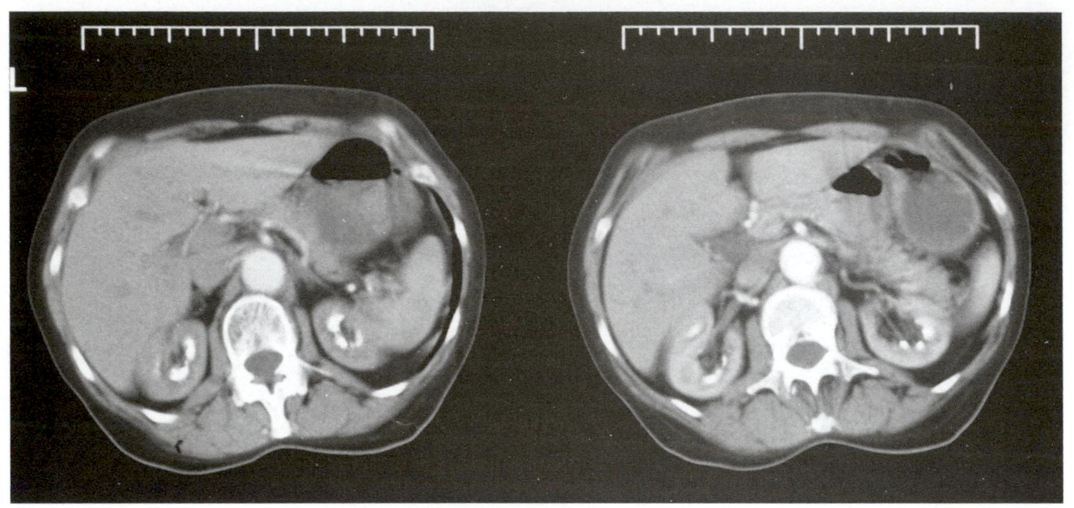

Note calcium deposits in the medulla of both kidneys. Courtesy of www.eMedicine.com

Questions

Name ______________________________ Section ________

1 What are the symptoms of kidney stones (nephrolithiasis)?

2 What is the etiology of nephrolithiasis?

3 How can the risk of future episodes of nephrolithiasis be reduced?

4 What are the symptoms of olecranon bursitis?

5 What is the etiology of olecranon bursitis?

6 What do you expect will be found in the bursal fluid?

7 What is causing his lower back pain now?

8 What are the risk factors for this current condition?

9 What are the treatments for this condition?

10 Why will Gerry M. need a kidney/liver transplant?

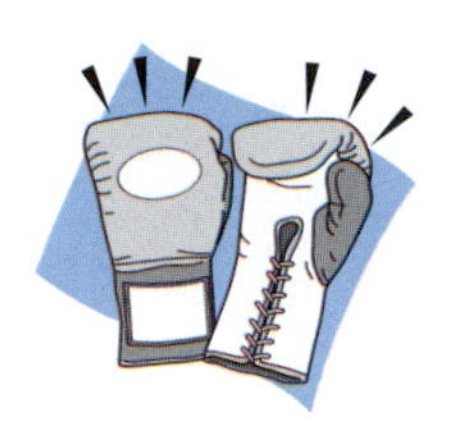

Answers to Questions

1 *What are the symptoms of kidney stones (nephrolithiasis)?*

A common symptom is severe pain over the costovertebral angle (which is over the kidney) that radiates to the lower abdomen and groin. The pain may last for minutes or hours, followed by relief if the stone is passed in the urine. Common associated symptoms include hematuria, fever, chills, nausea, and vomiting. Straining the urine with gauze to find stones can help confirm the diagnosis.

Region of Pain

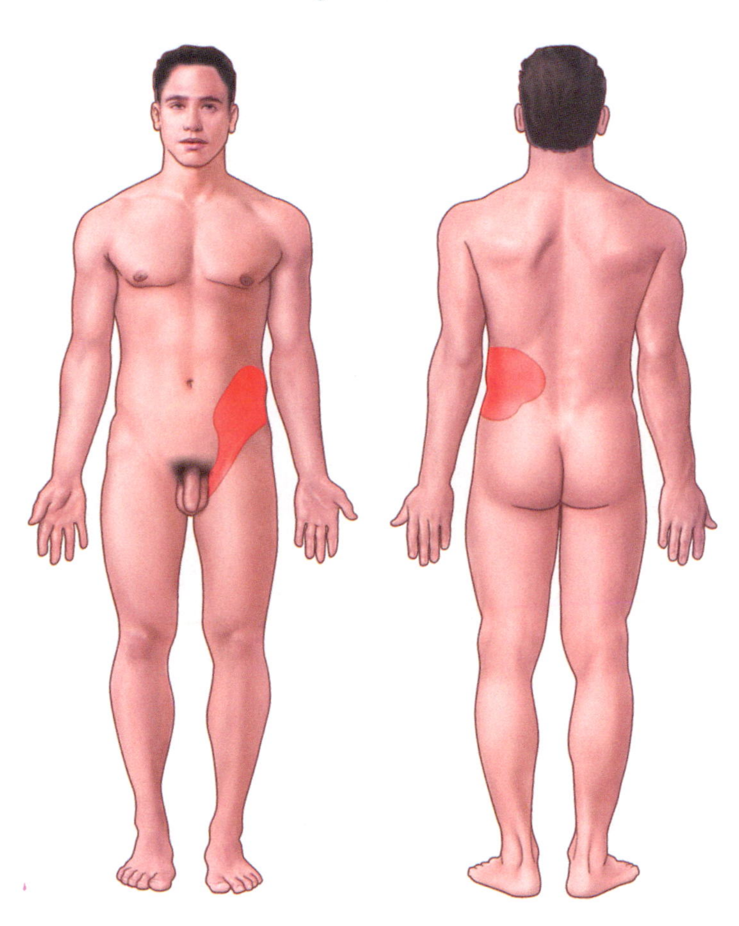

2 *What is the etiology of nephrolithiasis?*

Kidney stones can form in the minor and major calyces and can move to the ureter. Normally, the urine contains substances that prevent the formation of stones. (Additives in ice cream prevent ice crystals from forming in the same way.) If the concentration of certain solutes is high enough, the solutes can precipitate, forming crystals. These crystals can aggregate, forming stones. High blood levels of uric acid (as happens in gout) or oxalate can cause stone formation. Dehydration is often a factor because it can cause solutes to precipitate. Gerry M. was often dehydrated, as he laid carpet for hours on end and then went to the gym to work out. He was taking protein supplements (desiccated beef liver) that he bought from the gym. Dehydration and a high animal protein intake may have played a part in the formation of kidney stones.

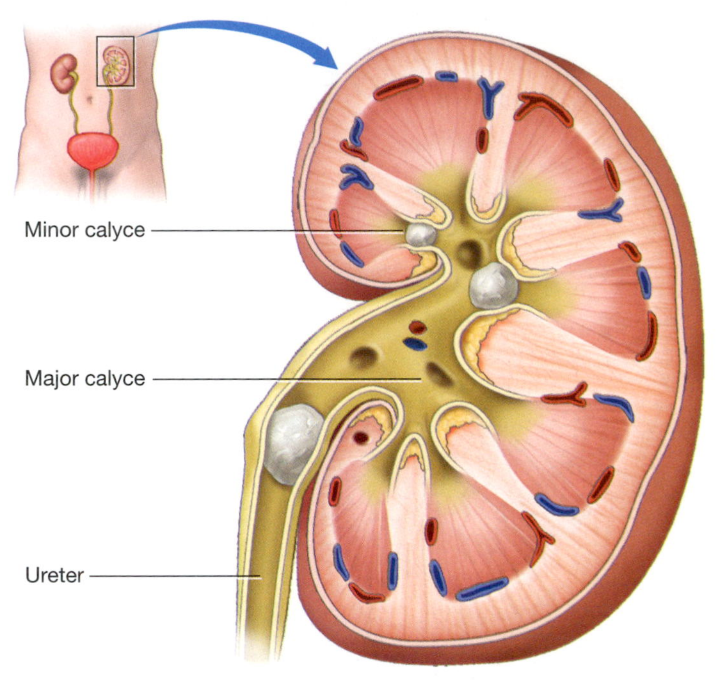

Calculi can lodge in the minor and major calyces of the kidney and the ureter.

3 *How can the risk of future episodes of nephrolithiasis be reduced?*

When Gerry M. was in the ED for the first episode of nephrolithiasis, his urine was strained and a small stone was captured. The stone was analyzed and found to be composed of calcium oxalate—the most common type of stone. The following evidence-based approaches have been shown to reduce the risk of calcium oxalate stones:

- Increase fluid intake (primarily from water) to 3–4 quarts per day such that the urine is a very pale yellow.
- Limit sodium intake to 2000 milligrams per day. (Sodium causes the kidneys to excrete more calcium.)
- Maintain a calcium intake that meets the RDA (recommended dietary allowance) for the patient's gender and age. (Low calcium diets do *not* work and can reduce bone mass.) Three servings of milk or cheese a day are advised.
- Reduce animal protein intake to no more than 4–6 ounces per day of beef, chicken, pork, fish, or eggs. (A serving the size of a deck of cards is about 3 oz.)
- Reduce intake of foods that are high in oxalate: peanuts and peanut butter, tea, instant coffee (more than 8 ounces a day), rhubarb, beets, beans, peas, and other legumes, green beans, berries such as blackberries, raspberries, strawberries, chocolate, Concord grapes, dark leafy greens such as spinach, beet tops, kale, Swiss chard and collards, oranges, tofu, sweet potatoes, and draft beer.
- Consume fruits and vegetables low in oxalate to maintain a low sodium-to-potassium ratio: asparagus, avocado, broccoli, Brussels sprouts, cauliflower, cabbage, cucumber, lettuce, sweet corn, turnips. (A low sodium intake combined with a high potassium intake is associated with a lower risk of calcium oxalate stones.)

- Avoid vitamin C supplements as well as calcium supplements and fish oil capsules, as these have been shown to increase stone formation in some people.
- Avoid non-steroidal anti-inflammatory agents (NSAIDs) including over-the-counter ibuprofen (Motrin, Advil) and naproxen (Aleve). NSAIDs reduce prostaglandins in the kidneys, which help to maintain renal blood flow. NSAIDs reduce oxalate and uric acid elimination by the kidneys.

4 *What are the symptoms of olecranon bursitis?*

Symptoms include pain and swelling over the olecranon. The patient has difficulty moving his elbow, and it is extremely painful to put the elbow on a flat surface. If the problem has been chronic, the bursal tissue may form thickened lumps that are tender and moveable.

5 *What is the etiology of olecranon bursitis?*

Olecranon bursitis can result from repetitive action (e.g., boxing or tennis) or from leaning the elbows on a hard surface (such as laying carpet or writing on a desk). The repeated injury causes the bursa to become irritated and thickened. Over time the chronic irritation can lead to acute inflammation and bursitis.

Olecranon bursitis also can occur following skin infection or from a spontaneous infection in the bursa. In addition to *Staphylococcal* and *Streptococcal* bacteria (skin flora), *Mycobacteria* (as in tuberculosis), *Brucella* and fungi can be infectious agents in bursitis.

Primary or secondary oxalosis and gout (hyperuricemia) also can cause bursitis because calcium oxalate or uric acid crystals can precipitate in the bursal fluid and cause inflammation.

6 *What do you expect will be found in the bursal fluid?*

Bursal fluid aspirated from his left elbow was analyzed. The WBC (white blood cell) count was normal, and no bacteria were found on Gram stain. Cultures were negative for bacteria including *Mycobacteria*, *Brucella*, and fungi. Bursal fluid was positive for calcium oxalate crystals. Gerry M.'s bursitis was caused by (a) repetitive action of boxing, (b) repetitively leaning his elbows on the floor while laying carpet, and (c) precipitation of calcium oxalate crystals in the bursa.

7 *What is causing his lower back pain now?*

Gerry M. is suffering from another episode of nephrolithiasis (kidney stones). He also has nephrocalcinosis, indicated by the calcium precipitates in the medulla of both kidneys. Most cases of nephrocalcinosis are asymptomatic. He also is now in acute renal failure (brought on, in part, from taking high-dose ibuprofen for the past 2 weeks). Unfortunately, on the first ED visit, Gerry M. did not report that his brother had died from renal failure. Gerry M. was only 10 at the time, so he does not know the actual cause of his brother's kidney disease.

The prescription-strength ibuprofen aggravated Gerry M.'s condition. Ibuprofen is a non-steroidal anti-inflammatory agent (NSAID) that reduces pain by inhibiting prostaglandins involved in inflammation. NSAIDs also reduce prostaglandins in the kidney, and this helps to maintain renal blood flow. The ibuprofen triggered acute renal failure. The ibuprofen also may have reduced the ability of Gerry M's kidneys to eliminate oxalate.

8 *What are the risk factors for this current condition?*

Nephrocalcinosis associated with calcium oxalate stones usually is a result of excess synthesis of oxalate by the liver or increased oxalate absorption in the intestine. Gerry M.'s high protein intake and dehydration also played a part in the formation of the calcium oxalate stones. He is given tests for genetic diseases that cause the liver to synthesize large amounts of oxalate. He is found to have an autosomal recessive disease called Primary Hyperoxaluria (PH) Type I, a result of insufficient synthesis of a hepatic enzyme called alanine-glyoxylate aminotrasferase (AGT). Deficiency of AGT leads to increased production of oxalate in the liver. Calcium oxalate also can precipitate in the heart and eyes. This is ruled out in Gerry M.'s case by an echocardiogram and eye exam.

9 *What are the treatments for this condition?*

If diagnosed early, before renal failure, individuals with PH can follow the dietary treatments indicated for calcium oxalate nephrolithiasis described above in the answer to question 3. Also, high doses of vitamin B_6 (pyridoxine) (1 g—which is more than 700% of the RDA of 1.3 mg) can help about half of patients with PH Type 1 by stimulating maximal AGT activity and reducing the amount of oxalate synthesized in the liver. Oxalate levels may even be reduced to normal levels in some patients. Magnesium supplements also help to reduce the ability of calcium to bind to oxalate.

10 *Why will Gerry M. need a kidney/liver transplant?*

Gerry M.'s PH was not diagnosed during his first visits to the ED for nephrolithiasis and bursitis. Unfortunately, the high-dose ibuprofen taken for 2 weeks contributed to his acute renal failure. Hemodialysis can remove oxalate, but not enough to prevent calcium oxalate from continuing to precipitate. Calcium oxalate would eventually precipitate in the heart, leading to death. A kidney/liver transplant can restore Gerry M. to health and essentially effect a cure for his PH. This was needed to save his life because his response to diet and vitamin B_6 was insufficient to lower his oxalate synthesis.

Later that month, a motorcyclist (not wearing a helmet) died in the hospital after hitting his head on a car door as it was opened on a busy street. The motorcyclist—a physical therapist with an organ donor card—was a perfect match for Gerry M. Unlike most patients with PH, who don't live long enough for a tissue match, Gerry M. was able to undergo a kidney/liver transplant. Today Gerry M. is manager of the carpet business and teaches cardio boxing. He also lectures to high school students about organ donor cards and the need to wear helmets.

References

Oxalosis and Hyperoxaluria Foundation
http://www.ohf.org/

Bursitis: Handuniversity.com
http://www.handuniversity.com/topics.asp?Topic_ID=3

Khan, A.N., & Macdonald, S. (2003). Nephrocalcinosis.
Accessed May 1, 2007 from eMedicine from *Web*MD. http://www.eMedicine.com/radio/topic470.htm

National Kidney Foundation. Kidney Stones.
http://www.kidney.org/atoz/atozItem.cfm?id=84

OrganDonor.Gov
http://www.organdonor.gov/

The National Transplant Society
http://www.organdonor.org/

10

The Red Hat Hikers

The Red Hat Hikers, a group of 60-something women, meet once a week to hike Mt. Monadnock, the second most climbed mountain in the world, after Mt. Fuji. On the schedule this week was the Pumpelly Trail, an arduous 4-hour climb with several hand and foot "scrambles" and a long view of the looming peak. All were ready with their hiking poles and camelback water bags. Sue was snack leader: "This week it's chocolate-covered dried cranberries, girls!"

"What—no gourp* this week?" exclaimed the other four women.

"Nope—these little things are supposed to be good for the bladder, and we can all use a little help there!"

The group soon settled into a stride where they could keep up a conversation in the humid 88-degree weather. When they got to the top, it was definitely a "summit day"—clear and cool, around 70 degrees. They munched their snack, drank water, and after about a 45-minute rest, started down the way they came, with fabulous views of the fall foliage to the north. As they descended, the air temperature got warmer and more humid.

"Temperature inversion!" Nan announced ". . . when the dense, cold air at the top traps the warm humid air at the bottom."

Sue had consumed all of her water and gasped. "Whoa—I just saw something weird, like a mountain lion, only it looked like a calico cat."

"Right," the others said, amused.

When the group reached the end of the bare rock above treeline, just before the last steep descent, Sue fell over backward.

"What's wrong Sue?" Nan quickly asked.

"I feel like my brain is shutting down. I can't find my words." Sue's voice trailed off.

Nan dialed 911, and a rescue team arrived in under 15 minutes. The medics put Sue on a mountain gurney and brought her down to a waiting ambulance.

*A loose mixture of salted nuts, chocolate chips, raisins, and sunflower seeds.

Selected lab values from Sue's admission to the ED are as follows:

Test	Value	Reference/Range (non-fasting)	Units
Glucose	115	<200	mg/dL
BUN	8	7–18	mg/dL
Creatinine, serum	0.9	0.8–1.3	mg/dL
Bilirubin, total	0.2	0.1–1.0	mg/dL
Protein, total	6.5	6.4–8.2	gm/dL
Albumin	3.4	3.4–5.0	gm/dL
Alkaline phosphatase	130	50–136	U/L
AST/SGOT	29	15–37	U/L
ALT/SGOT	41	30–65	U/L
GGT	39	1–94	U/L
Cholesterol, total	180	50–199	mg/dL
Triglycerides	75	15–149	mg/dL
WBC	5.1	4.5–11.0	K/cmm
RBC	4.0	4.00–5.20	K/cmm
HGB	11.9	12.0–16.0	g/dL
HCT	35.8	36.0–46.0	%
MCV	90	80–100	fL
MCH	31.2	26.0–34.0	pg
MCHC	33.7	31.0–37.0	g/dL
PT (prothrombin time)	11	9–12	seconds
Sodium	129	136–145	mmol/L
Potassium	4.5	3.5–5.1	mmol/L
Chloride	97	98–107	mmol/L
CO_2, Total	26.3	21.0–32.0	mmol/L
Anion Gap	10.7	8.0–17.0	mEq/L
Calcium	8.5	8.5–10.5	mg/dL

Questions

Name ______________________________ Section __________

1 Which lab values are abnormal?

2 What problem does Sue have?

3 What is the most likely cause of this problem?

4 What are physical symptoms of this problem?

5 How could this problem have been prevented?

6 How can this problem be treated?

Answers to Questions

1 *Which lab values are abnormal?*

Sue's sodium and chloride are low. Her hemoglobin and hematocrit are low as well.

2 *What problem does Sue have?*

Sue has hyponatremia. Her mean corpuscular hemoglobin (MCH) and mean corpuscular hemoglobin concentration (MCHC) and mean corpuscular volume (MCV) are all normal, suggesting that her red blood cells are normal. Her blood urea nitrogen (BUN), creatinine, bilirubin, protein, and albumin are all on the low normal side. Her lab values, taken together with her history of hiking in hot, humid weather while drinking large amounts of water suggest that Sue has dilutional hyponatremia—a low sodium concentration resulting from copious water consumption without adequate sodium replacement. She is also hypervolemic.

3 *What is the most likely cause of this problem?*

Sue was hiking on a warm, humid day and likely perspired. She drank water and ate chocolate-covered dried cranberries but had an insufficient sodium intake to replace the losses. She drank a lot of water—perhaps more than needed to replace the water loss. This resulted in water intoxication with dilution of her blood cells and electrolytes. Her potassium remained in the normal range, perhaps because chocolate-covered cranberries are high in potassium.

Another contributor to Sue's case might be the "syndrome of inappropriate antidiuretic hormone," or SIADH. SIADH causes an increase in free water without change in total body sodium because of increased vasopression secretion. SIADH is more common in older adults and also is related to thiazide diuretics taken for hypertension and selective serotonin reuptake inhibitors (SSRIs) taken for depression. Sue has been taking Zoloft, an SSRI, for 5 years.

Low sodium concentrations in the blood cause water to move from the blood to the cells in an attempt to normalize sodium concentrations. Most cells can tolerate the resulting swelling, but brain cells cannot do so because the skull encases the brain.

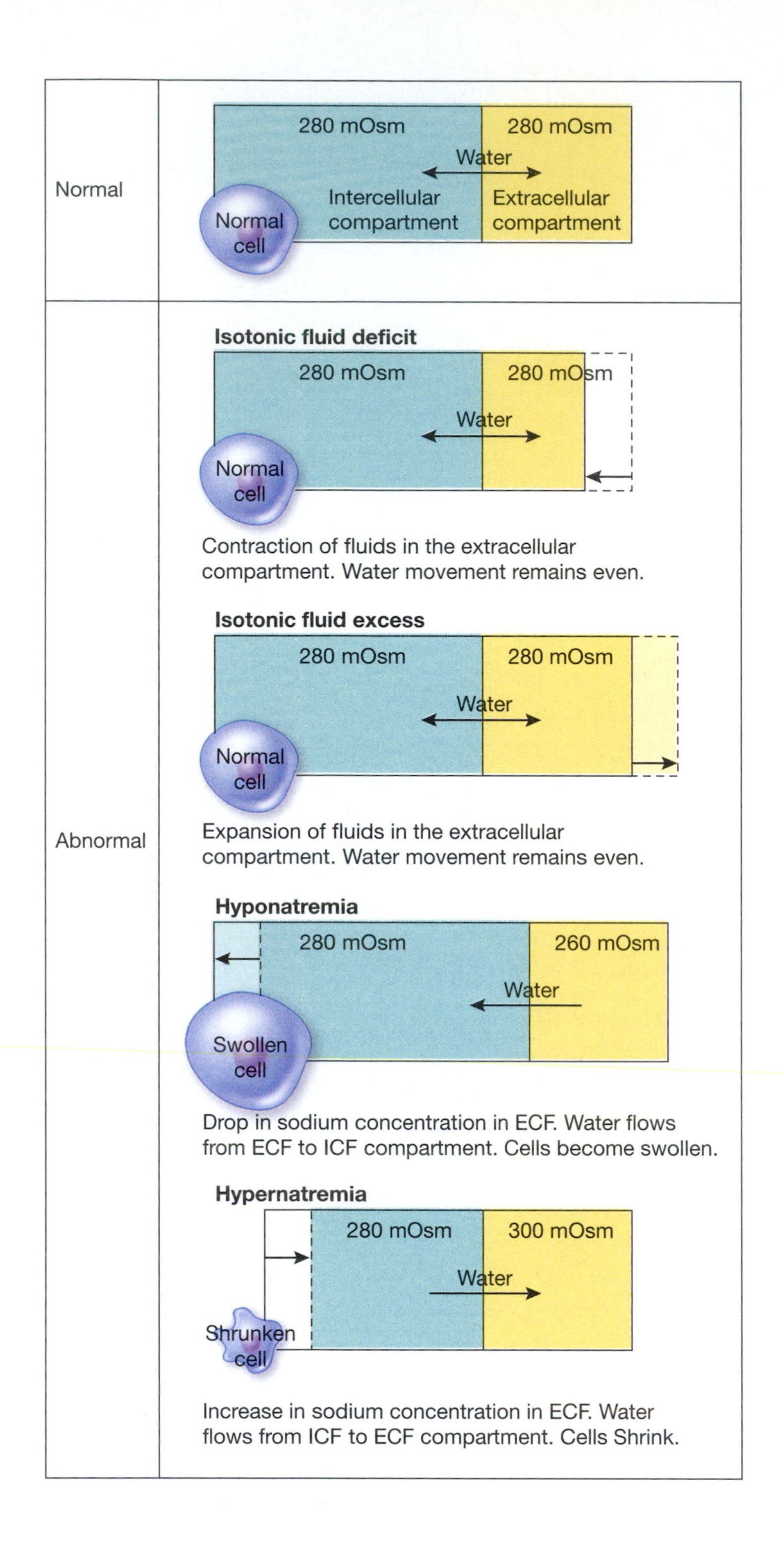
Normal
280 mOsm
280 mOsm
Water
Intercellular compartment
Extracellular compartment
Normal cell
Abnormal
Isotonic fluid deficit
280 mOsm
280 mOsm
Water
Normal cell
Contraction of fluids in the extracellular compartment. Water movement remains even.
Isotonic fluid excess
280 mOsm
280 mOsm
Water
Normal cell
Expansion of fluids in the extracellular compartment. Water movement remains even.
Hyponatremia
280 mOsm
260 mOsm
Water
Swollen cell
Drop in sodium concentration in ECF. Water flows from ECF to ICF compartment. Cells become swollen.
Hypernatremia
280 mOsm
300 mOsm
Water
Shrunken cell
Increase in sodium concentration in ECF. Water flows from ICF to ECF compartment. Cells Shrink.

4 *What are physical symptoms of this problem?*

Most of the symptoms of dilutional hyponatremia are caused by swelling of brain cells. Abnormal mental status with confusion and hallucinations are most common with the sodium levels that Sue is experiencing. Loss of appetite, headache, nausea, vomiting, muscle weakness, and muscle cramps also occur. Without treatment, decreased consciousness and coma may result.

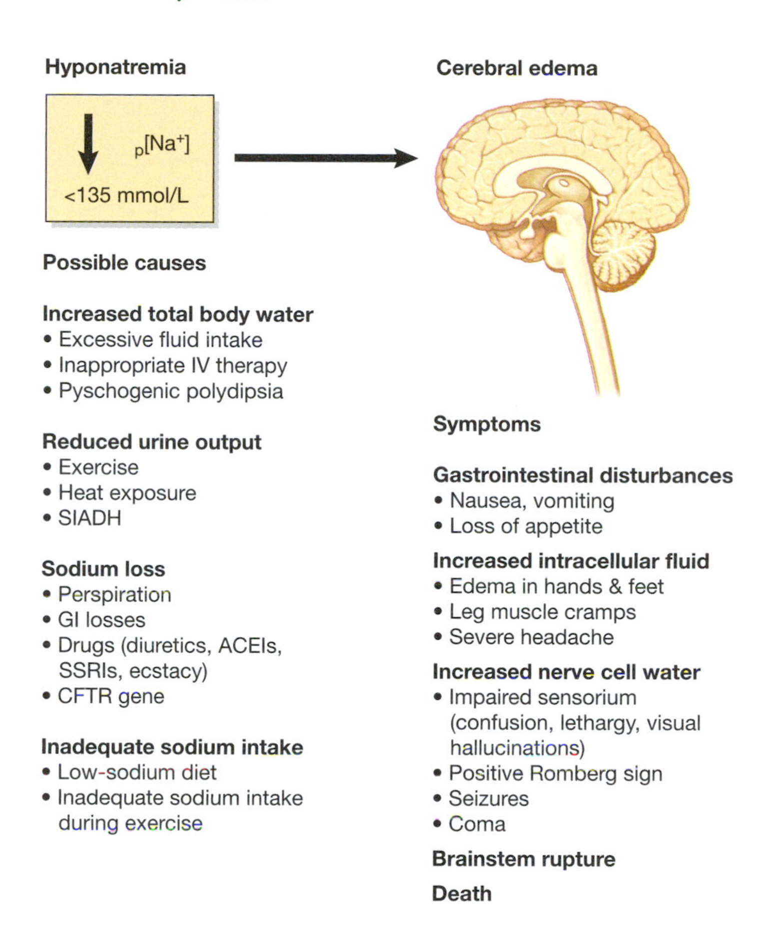

5 *How could this problem have been prevented?*

Eating a hiking snack such as "gourp" would have contributed sufficient sodium to replace her losses. Salt tablets are *not* recommended, or necessary. Also, use of a "camelback type" water bag can lead to unintentional water overload as people tend to sip water constantly as they hike. They are not able to assess how much water they have consumed as they would by drinking from a water bottle.

6 *How can this problem be treated?*

A rapid increase in sodium levels could increase the risk of central pontine myelinolysis (CPM). CPM is a destruction of the myelin sheath due to a rapid change in sodium levels in the blood, most often when hyponatremia is corrected too quickly. CPM causes long-term nerve damage, resulting in double vision, difficulty walking, and reduced muscle strength.

Sue is hypervolemic, so water restriction with a diet high in sodium is ordered. In 24 hours Sue is discharged on her "normal" diet. She is ready to become the "snack leader": "Super gourp and pretzels next week, girls! And I'm going back to carrying water bottles in my pack. "If it ain't broke, don't fix it. I don't think I need these new-fangled things!"

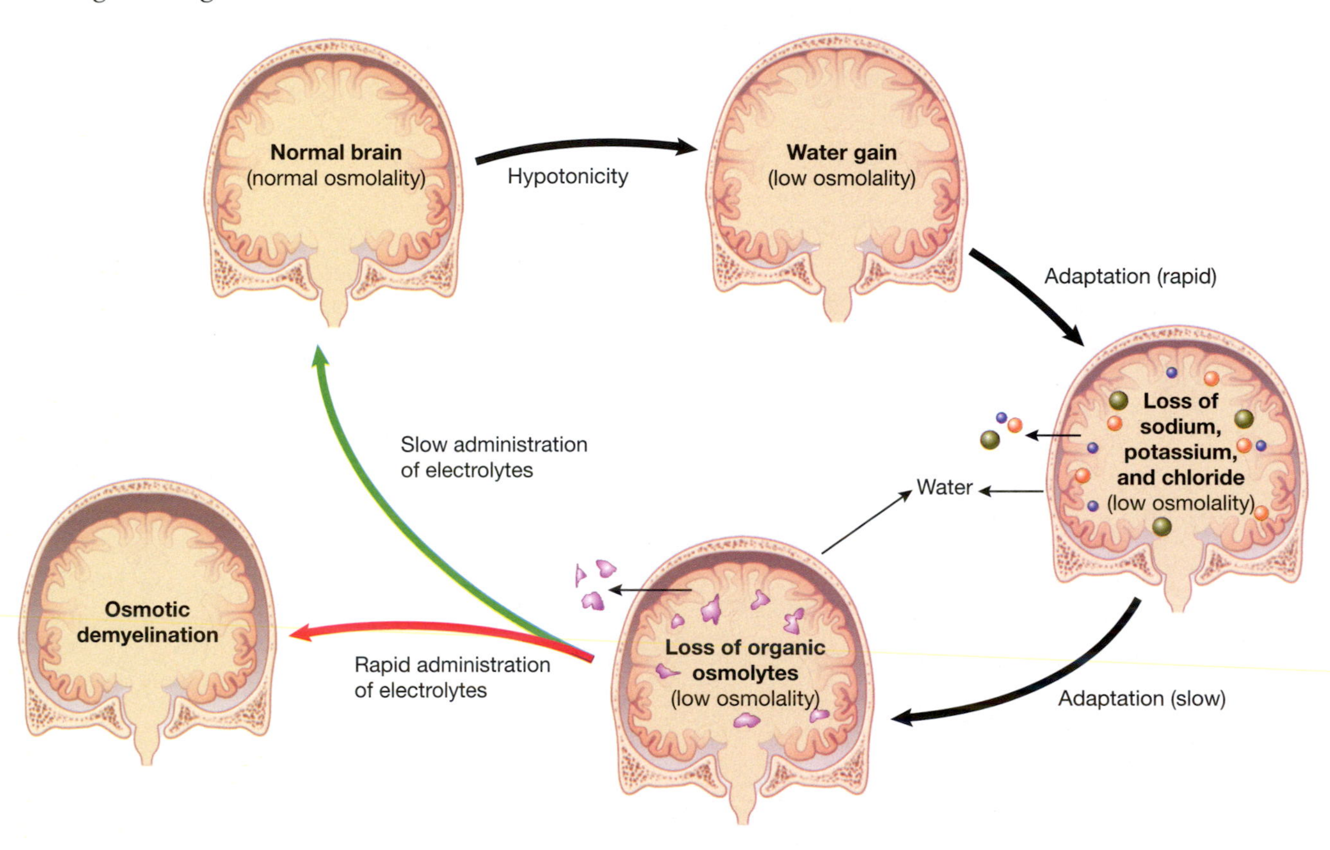

References

Risks of overhydration with exercise.
http://www.cptips.com/water.htm

Luzzio, C.L. (2007). Central pontin myelinolysis. Accessed January 9, 2009 from eMedicine from *Web*MD.
http://www.eMedicine.com/neuro/topic50.htm

Eat to Compete. Iowa State University Extension.
http://www.extension.iastate.edu/nutrition/sport/fluids.html

Murray, B., Stofann J., Eichner, E.R. (2003). Hyponatremia in athletes. Sports Science Exchange 88, 16(1).
http://www.gssiweb.com/Article_Detail.aspx?articleID=604

Neafsey, P.J. (2004). Thiazides and selective serotonin reuptake inhibitors can induce hyponatremia. *Home Healthcare Nurse*, 22(11), 788–790.

11

The Quad Riders

Cheryl and Sean enjoy riding their quads (ATVs) at night on the back roads under the full moon and stars. One night Cheryl bumped over a hole and fell off her quad next to a stone wall. Sean could not avoid running over her and hitting the stone wall. He fell off his quad, hit his head, and became unconscious. His quad landed on Cheryl, crushing both of her legs. She was unable to reach her cell phone to call for help.

Nearly 4 hours later, Sean "came to" and was able to dial 911. He pulled the quad off Cheryl. The LifeStar helicopter arrived within 15 minutes. "Wish you hadn't pulled the quad off your wife yet," said the paramedic. "We want to hook her up to a cardiac monitor, give her albuterol via a nebulizer and hook her up to a fluid IV before we relieve the crush."

"But she doesn't even have any bruises or swelling," Sean responded. "And she didn't have any pain until *after* I removed the quad."

The paramedics did their work. They put in two IVs of normal saline at 500 ml/hour total. They also gave Cheryl a dose of sodium bicarbonate (in the IV bag), and fentanyl (an opioid) for pain. Then LifeStar transported both Cheryl and Sean to the nearest trauma center, a 15-minute chopper ride away.

Sean was diagnosed with a severe concussion. His vital signs and results of an MRI were within normal limits. He was admitted for observation.

Cheryl suffered crush injury syndrome, which results from a severe crush injury over an extensive portion of the body that lasts for a prolonged period, typically 4–6 hours of compression.

Following are serial ECGs—one taken immediately after the paramedics arrived, and one 5 minutes later in the chopper.

Normal ECG

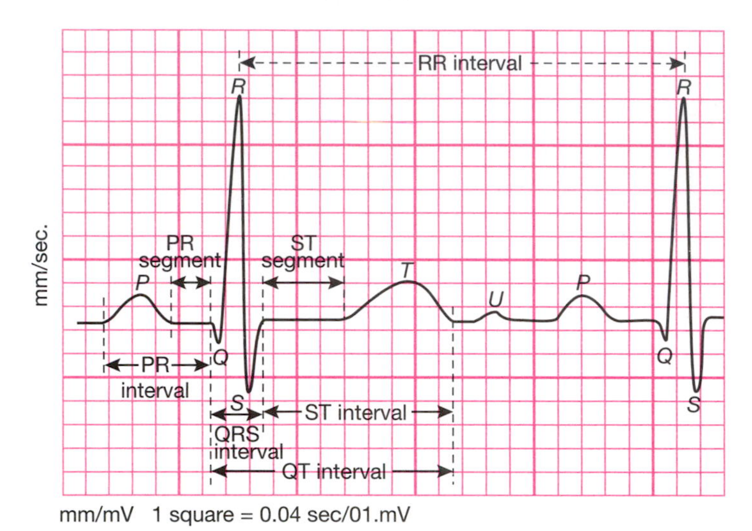

Cheryl's Initial ECG

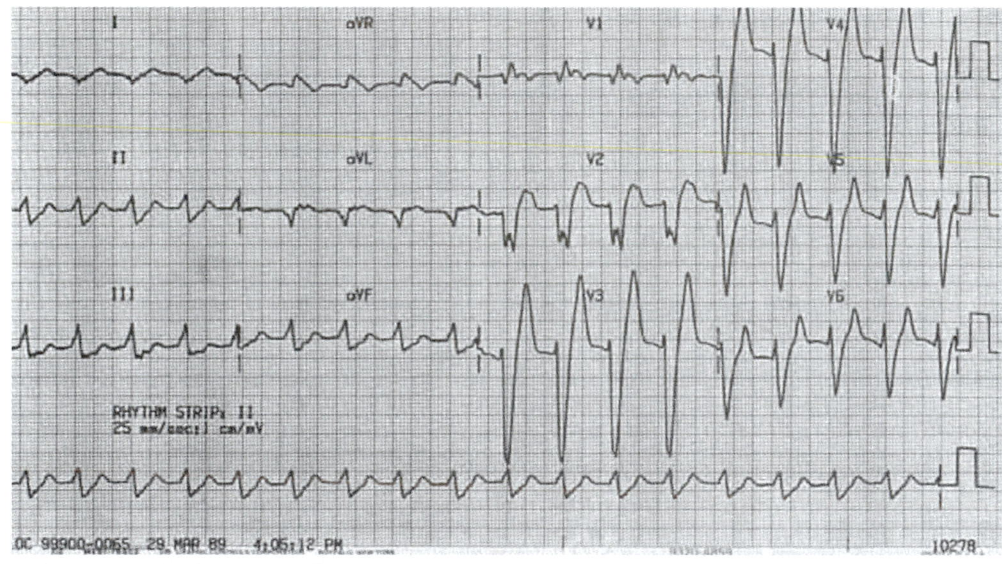

Note widened QRS complexes, flat and wide P waves, ventricular ectopic beats; serum potassium is probably ~ 7.5–8.5 mEq/L.

Image reprinted with permission from www.eMedicine.com, 2008

Cheryl's ECG 5 Minutes After IV of Normal Saline and Dose of Sodium Bicarbonate

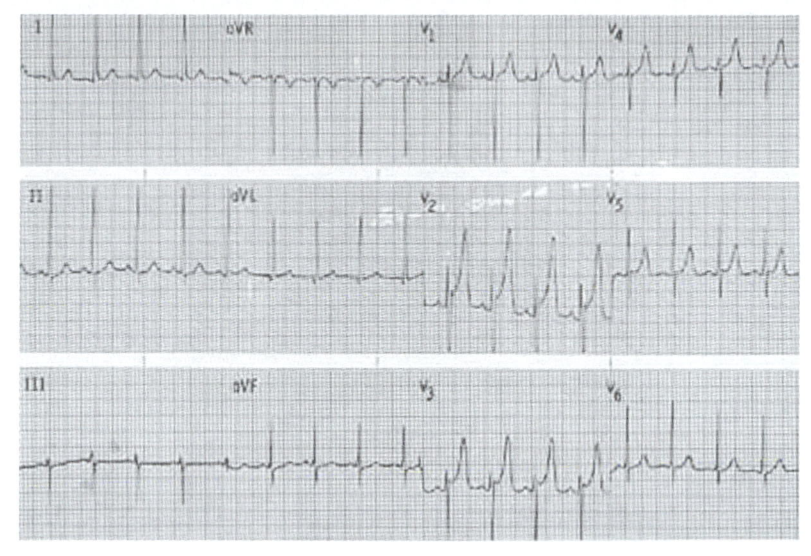

Note improvement: QRS less wide; there are peaked T waves; serum potassium is ~ 5.5–6.5 mEq/L.

Image reprinted with permission from www.eMedicine.com, 2008

When Cheryl arrived at the trauma center, her hands were in an unusual position:

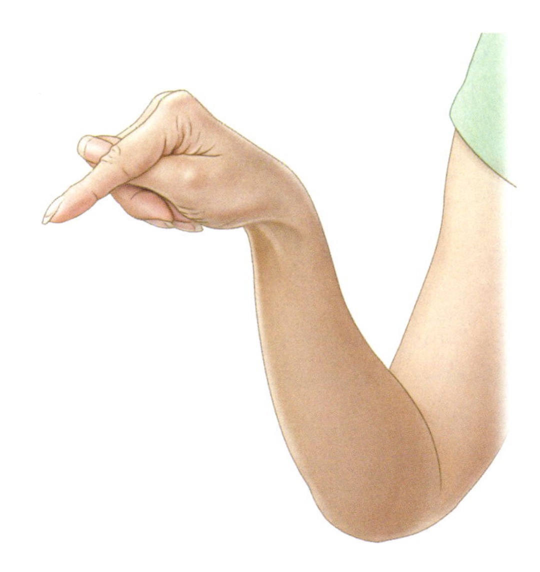

Her ECG follows:

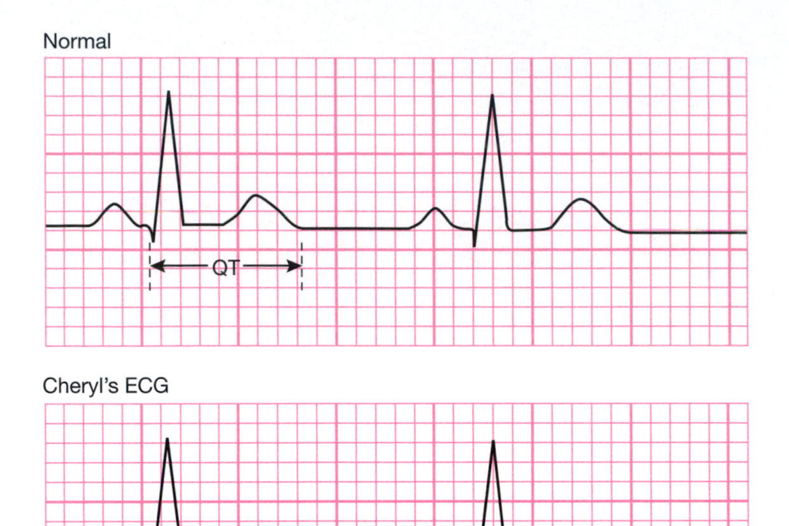

Cheryl had a Foley catheter inserted, and mannitol was added to her IV infusion to maintain urine output. Her urine had a dark, red-brown color. She was given an infusion of glucose and insulin.

Below are selected lab values for Cheryl 5 minutes later.

Test	Value	Reference Range (non-fasting)	Units
Glucose	115	<200	mg/dL
BUN	18	7–18	mg/dL
Creatinine, serum	1.9	0.8–1.3	mg/dL
Bilirubin, total	1.2	0.1–1.0	mg/dL
Protein, total	6.5	6.4–8.2	gm/dL
Albumin	3.4	3.4–5.0	gm/dL
Creatine Phosphokinase (CPK)	80,000	8–150	IU/L
WBC	5.1	4.5–11.0	K/cmm
RBC	4.0	4.00–5.20	K/cmm
HGB	11.9	12.0–16.0	g/dL
HCT	35.8	36.0–46.0	%
MCV	90	80–100	fL
MCH	31.2	26.0–34.0	pg
MCHC	33.7	31.0–37.0	g/dL
Sodium	140	136–145	mmol/L
Potassium	5.5	3.5–5.1	mmol/L
Chloride	108	98–107	mmol/L
Calcium	8.0	8.5–10.5	mmol/L
Phosphate	4.9	2.5–4.5	mg/dL

One of Cheryl's IV lines was removed and carefully flushed with sterile water. Then Cheryl was given calcium chloride by slow IV push (over a period of 60 seconds).

Questions

Name ______________________________ Section __________

1 Why did the paramedic state that the cardiac monitor, IV fluid line, and albuterol should be administered *before* removing the quad?

2 Albuterol is a bronchodilator. What is released in crush injury syndrome that would cause bronchoconstriction?

3 What toxic substances released from crushed muscle can affect the heart? (*Hint:* Consider two different minerals released from crushed muscle.)

4 What toxic substances released from crushed muscle can affect the kidneys?

5 The paramedics put in two IV lines of normal saline to counteract the expected hypovolemia and prevent hypovolemic shock. Why did they also administer sodium bicarbonate?

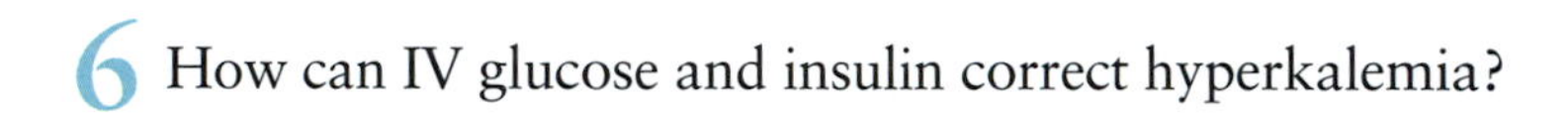

6 How can IV glucose and insulin correct hyperkalemia?

7 What does Cheryl's cramped hand and lab values suggest?

8 Why can't calcium chloride be put in the same infusion with sodium bicarbonate?

Answers to Questions

1 *Why did the paramedic state that the cardiac monitor, IV fluid line, and albuterol should be administered before removing the quad?*

The crushing injury damages muscle and blood vessels and causes release of numerous toxic substances. While the object is still crushing the muscle, these substances cannot reach the general circulation. Once the crushing object is removed, however, the patient will be subjected to immediate hypovolemia from loss of intravascular fluids leaking through cell membranes and capillaries. In addition, the crushed muscle cells release toxins that can continue for as long as 60 hours after release of the crushing force. Ideally the patient is treated before the object is removed.

2 *Albuterol is a bronchodilator. What is released in crush injury syndrome that would cause bronchoconstriction?*

When muscle cells are crushed and then the force is removed, inflammatory mediators are released in large numbers. These inflammatory mediators cause large numbers of mast cells to degranulate and release histamine and leukotrienes, which are bronchoconstrictors. Leukotrienes can also damage the lungs and liver. Prostaglandins released from mast cells can damage the lungs as well.

3 *What toxic substances released from crushed muscle can affect the heart?*

Most of the potassium in the body is intracellular. When large numbers of muscle cells are crushed, potassium leaks out and causes hyperkalemia. Hyperkalemia causes cardiac dysrhythmias—especially when acidosis (low blood pH) and hypocalcemia are also present. Muscle cells that have been crushed for 4 or more hours contain a lot of lactic acid because available oxygen has been insufficient to complete the metabolic process and break down lactic acid into carbon dioxide and water. As a result, patients develop metabolic acidosis.

Crushed muscle also releases phosphate, resulting in hyperphosphatemia. The elevated blood phosphate levels will precipitate some serum calcium and lead to hypocalcemia. Hypocalcemia causes cardiac dysrhythmias.

4 *What toxic substances released from crushed muscle can affect the kidneys?*

The crushed muscle cells release large amounts of myoglobin. The myoglobin can precipitate in the kidneys. Called rhabdomyolysis, this can cause acute kidney injury (AKI). AKI is more likely if the urinary pH is low. The dark brown/red color of Cheryl's urine results from myoglobin.

5 *The paramedics put in two IV lines of normal saline to counteract the expected hypovolemia and prevent hypovolemic shock. Why did they also administer sodium bicarbonate?*

After the weight was removed from the crushed muscle cells, they released cell contents, including acids such as lactic acid, which built up in the muscle but could not be further metabolized because oxygen was not delivered to complete the metabolic process. Additional acids that are released include amino acids and other organic acids. The sodium bicarbonate helps to maintain a normal blood pH and counteract acidosis.

Correcting blood pH with sodium bicarbonate also can help protect the heart from hyperkalemia. Hyperkalemia is more dangerous for the heart in the presence of acidosis.

The sodium bicarbonate also will increase the pH of the urine, and this will help to reduce the amount of myoglobin that will precipitate in the kidneys.

6 *How can IV glucose and insulin correct hyperkalemia?*

Insulin interacts with insulin receptors on muscle and fat cells and drives glucose and potassium into cells. This process is one of the reasons that most potassium is intracellular. Giving glucose and insulin will lower Cheryl's potassium levels and help protect her heart from hyperkalemia. She will need to have her glucose levels monitored carefully for possible *hypo*glycemia. *Hyper*glycemia is also possible because trauma increases the release of cortisol from the adrenal glands, growth hormone from the pituitary, and glucagon from the pancreas. These hormones will stimulate liver synthesis of glucose through a process of gluconeogenesis. The new glucose is made from amino acids (some of which are from the crushed muscle) and from glycerol on triglycerides released from fat.

7 *What does Cheryl's cramped hand and lab values suggest?*

Cheryl developed hyperphosphatemia (resulting from release of phosphate from crushed muscle). The phosphate precipitated with some serum calcium and caused hypocalcemic tetany by the time she reached the trauma unit. She also has elevated CPK because of release of the enzyme from crushed muscle cells. She still has mild hyperkalemia.

8 *Why can't calcium chloride be put in the same infusion with sodium bicarbonate?*

The calcium will precipitate as calcium carbonate (chalk/limestone). This could clog both the IV line and Cheryl's vein. Calcium chloride should *never* be administered with sodium bicarbonate. Cheryl's repeat blood tests show that the improvement in her serum calcium is holding. This is good news, as often hypocalcemia returns. A constant calcium carbonate infusion is dangerous in patients with crush injury syndrome because the calcium can calcify the injured muscle and aggravate rhabdomyolysis.

Both Cheryl and Sean fully recover, thanks to prompt trauma care. As she is about to be discharged, news on the TV of a major earthquake prompts Sean to say, "I wish everyone knew what I know now—always get medical help to a person who is being crushed *before* you remove the object. Cheryl, you would have been better off if I had just waited 15 minutes."

Cheryl responded, "Hon, we didn't know. But let's go out and get those helmets. That is something we *did* know and didn't act on. Your knock on the head that put you out for 4 hours is what caused all this. And Hon—let's not ride at night anymore, okay?"

References

Krost, W.S., Mistovich, J.J., & Limmer, D. (2008). Beyond the basics: crush injuries and compartment syndrome. Emergency Medical Services.
http://www.emsresponder.com/print/Emergency--Medical-Services/Beyond-the-Basics--Crush-Injuries-and-Compartment-Syndrome/1$7056

Garth, D. (2006). Hyperkalemia. *eMedicine*, April 25
http://www.eMedicine.com/emerg/topic261.htm

Pegoraro, A.A., & Rutecki, G.W. (2007). Hypocalcemia. *eMedicine*, Feb. 1.
http://www.eMedicine.com/med/topic1118.htm

Vanholder, R., van der Tol, A., De Smet, M., Hoste, E., Koc, M., Hussain, A., Khan, S., & Sever, M.S. (2007). Earthquakes and crush syndrome casualties: Lessons learned from the Kashmir disaster. *Kidney International, 71*, 17–23.
http://www.nature.com/ki/journal/v71/n1/full/5001956a.html

12

Stargazed

Mrs. D., age 78, has been living at Stargaze Healthcare Center for 2 months following hospitalization for a broken hip. She is frail and is confined to bed and wheelchair. She is moved to the wheelchair by use of a lift. Her medications include insulin for Type 2 diabetes, lactulose for constipation, Percocet for pain as needed (contains acetaminophen as in Tylenol and oxycodone, an opioid), and Xanax as needed for anxiety. Nursing student John Smart has been observing the care of Mrs. D. and has found the following, which he writes in his clinical course journal:

> There is no documentation of her being repositioned every 2 hours per protocol. Her chart reveals that she has been getting a Percocet at 11 a.m. and 5 p.m. on Mondays, Tuesdays, Wednesdays, and Thursdays for 'pain in the lower back.' She has been getting Xanax at 4 p.m. on Mondays, Tuesdays, Wednesdays, and Thursdays for 'agitation and anxiety.' Her blood glucose is checked daily at 7 a.m. and 4 p.m. Blood glucose levels are typically 130–140 at 7 a.m. and 210–280 at 4 p.m. The chart states that she is 'incontinent in both bladder and bowels' and she is kept in adult diapers.

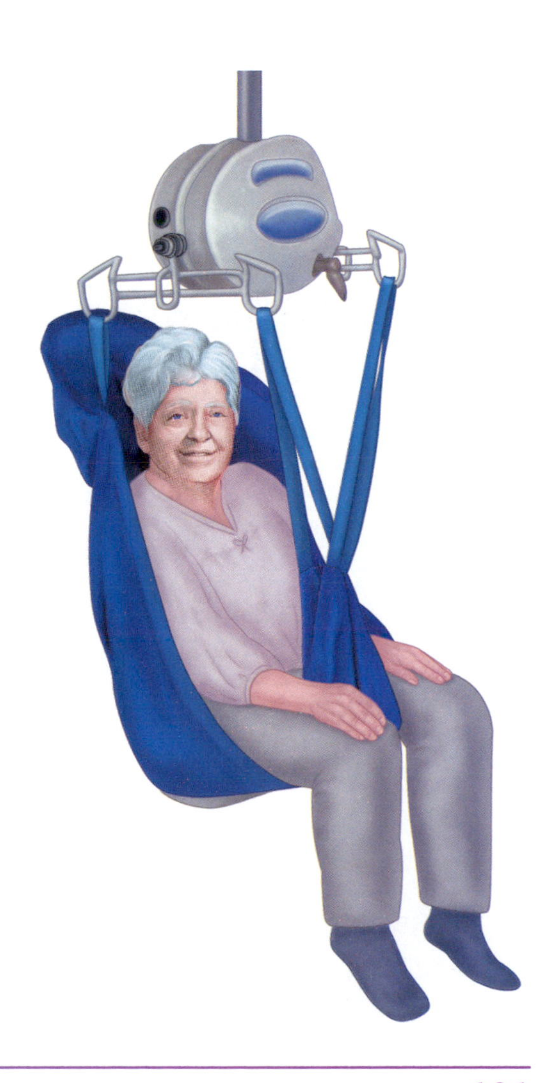

John observes that during the week, Mrs. D. has severe short-term memory problems but seems better on the weekends. For example, when the staff came in with the patient lift last Tuesday and Thursday during his clinical, Mrs. D. could not remember what the lift was for and "she was anxious as if she was seeing it for the first time." When he stopped in on Sunday to review her chart, he observed the staff placing Mrs. D. in the lift while she was calm and chatting. She spotted John and called out, "Hello, young man! Nice to see you again! They've come to take me to the toilet. Talk to you later!"

Questions

Name ____ Section __________

1 Why do you think Mrs. D. does *not* have short-term memory loss on the weekends? (*Hint:* Benzodiazepine sedatives, including Xanax, cause anterograde amnesia.)

2 Lactulose is a synthetic sugar that is not absorbed. It increases the osmolality of the stool. Osmolality is the concentration of osmotically active particles that attract water. The increased stool osmolality reduces water absorption from the colon so more is left in the stool, thereby promoting laxation. What possible side-effect do you think this could cause?

3 Do you think Mrs. D. is really incontinent? Why or why not? (*Hint*: Consider *when* the Percocet is given for "pain in the lower back.")

4 What physical assessment results and lab values would you be interested in looking at?

A blood sample was taken on Friday at 4 p.m. Mrs. D. was catheterized so a 24-hr urine sample could be analyzed.

Mrs. D's selected lab values follow:

Test	Value	Reference Range (non-fasting 4 p.m.)	Units
Glucose	240	<200	mg/dL
BUN	18	7–18	mg/dL
Creatinine, serum	1.9	0.8–1.3	mg/dL
Bilirubin, total	1.2	0.1–1.0	mg/dL
Protein, total	7.5	6.4–8.2	gm/dL
Albumin	4.4	3.4–5.0	gm/dL
WBC	6.1	4.5–11.0	K/cmm
RBC	5.5	4.00–5.20	K/cmm
HGB	15.2	12.0–16.0	g/dL
HCT	48.8	36.0–46.0	%
MCV	90	80–100	fL
MCH	31.2	26.0–34.0	pg
MCHC	33.7	31.0–37.0	g/dL
Sodium	148	136–145	mmol/L
Potassium	4.9	3.5–5.1	mmol/L
Chloride	110	98–107	mmol/L
Calcium	10.0	8.5–10.5	mg/dL

Following are the results of her urine test.

Test	Value	Reference Range (non-fasting 4 p.m.)	Units
Urine specific gravity	1.09	1.002–1.028	
Urine osmolality	1447	50–1400	mOsm/kg
Glucose	absent		
Microalbumin	absent		
Protein	absent		
Sodium	347	15–250	mEq/24 h

5 What do Mrs. D.'s blood and urine lab values suggest?

6 Why didn't Mrs. D. complain of thirst?

7 How will Mrs. D. be treated?

Answers to Questions

1 *Why do you think Mrs. D. does not have short-term memory loss on the weekends?*

Anterograde amnesia is the inability to form memories after an incident. In Mrs. D.'s case, Xanax may help reduce her anxiety about the patient lift, but she may forget how the lift works the next day. She is not getting Xanax on Friday through Sunday, so her memory is better on the weekend days.

2 *Lactulose is a synthetic sugar that is not absorbed. It increases the osmolality of the stool. Osmolality is the concentration of osmotically active particles that attract water. The increased stool osmolality reduces water absorption from the colon so more is left in the stool, thereby promoting laxation. What possible side-effect do you think this could cause?*

The majority of water absorption occurs in the colon. If Mrs. D. does not drink enough fluids, she may become dehydrated because the lactulose will cause less water absorption in the colon.

3 *Do you think Mrs. D. is really incontinent? Why or why not?*

The chart states that Mrs. D. requires pain relief for her lower back only on Mondays, Tuesdays, Wednesdays, and Thursdays but not on other days of the week. This suggests that there is a trigger for her pain on Monday through Thursday. Perhaps she is left for hours in her wheelchair and is not taken to the toilet. This would result in lower back pain and soiling. The assessment of incontinence is likely inappropriate. The fact that she is able to articulate that "they are taking me to the toilet" on a Sunday supports this conclusion.

4 *What physical assessment results and lab values would you be interested in looking at?*

Assessment of skin turgor (for dehydration) and electrolyte concentrations in the blood may identify dehydration. Many older people have poor skin turgor because of loss of skin elasticity, so skin turgor by itself may be a poor indicator of hydration. In Mrs. D.'s case, however, it provided a helpful clue! Careful assessment of fluid intake and output (I&O) may indicate that Mrs. D.'s fluid intake is less than her output. An increased urine-specific gravity, sodium, and osmolality also suggest dehydration, as does tachycardia with a weak and thready pulse.

A physical assessment of Mrs. D. on Monday morning suggests that she is well hydrated. Her skin has good turgor, and she has no symptoms of dehydration. Her pulse is 88 and strong. On Friday, however, Mrs. D. has poor skin turgor, and a weak and thready pulse of 99, indicating dehydration.

The staff had not been assessing her fluid output because Mrs. D. is kept in adult diapers. Mrs. D. has a water mug by her bedside. The water is changed every 4 hours between 8 a.m. and 4 p.m. (three times a day, 24 oz or 710 cc) and the amount remaining is measured. The record for Monday—Friday suggests that no water was left in the water mug. Taken together with her fluid intake at meals, it was calculated that Mrs. D. was taking in 2000 ccs of fluid per day, which is adequate for her age and consideration of lactulose medication.

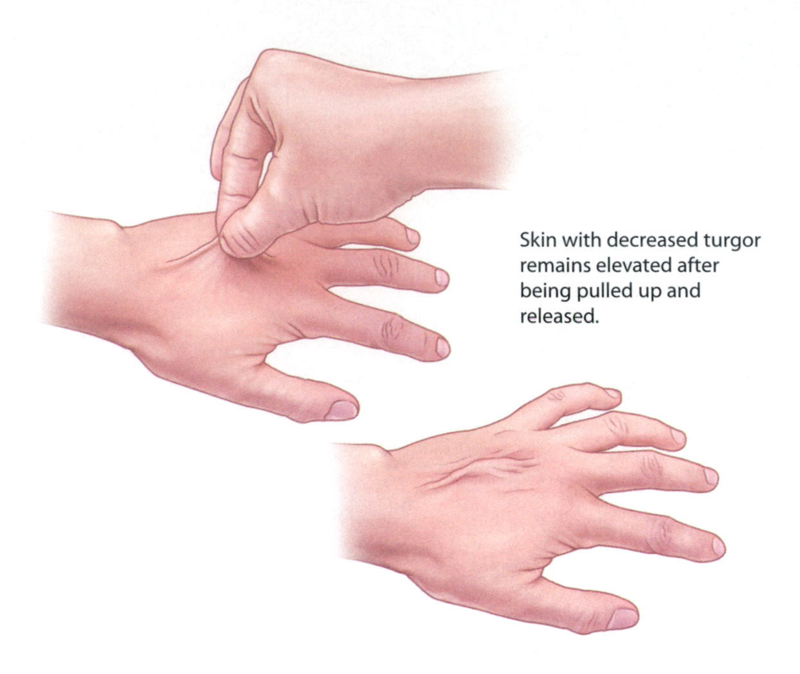

Her stools are reported to be formed and not watery. Her weight, however, was recorded on Thursday as 107 pounds, which is 5 pounds less than her weight on Monday when John Smith weighed her. He noted that her pulse was 98 and weak and thready on Thursday but 88 and strong on Monday. One pound of acute weight loss = 500 cc of fluid.

A blood sample was taken on Friday at 4 p.m. Mrs. D. was catheterized so a 24-hr urine sample could be analyzed.

Mrs. D.'s selected lab values:

Test	Value	Reference Range (non-fasting 4 p.m.)	Units
Glucose	240	<200	mg/dL
BUN	18	7–18	mg/dL
Creatinine, serum	1.9	0.8–1.3	mg/dL
Bilirubin, total	1.2	0.1–1.0	mg/dL
Protein, total	7.5	6.4–8.2	gm/dL
Albumin	4.4	3.4–5.0	gm/dL
WBC	5.1	4.5–11.0	K/cmm
RBC	5.5	4.00–5.20	K/cmm
HGB	15.2	12.0–16.0	g/dL
HCT	48.8	36.0–46.0	%
MCV	90	80–100	fL
MCH	31.2	26.0–34.0	pg
MCHC	33.7	31.0–37.0	g/dL
Sodium	148	136–145	mmol/L
Potassium	4.9	3.5–5.1	mmol/L
Chloride	110	98–107	mmol/L
Calcium	10.0	8.5–10.5	mg/dL

Following are results of her urine test.

Test	Value	Reference Range (non-fasting 4 p.m.)	Units
Urine-specific gravity	1.09	1.002–1.028	
Urine osmolality	1447	50–1400	mOsm/kg
Glucose	absent		
Microalbumin	absent		
Protein	absent		
Sodium	347	15–250	mEq/24 h

5 *What do Mrs. D.'s blood and urine lab values suggest?*

Mrs. D. has apparent hemoconcentration as she has elevated serum glucose, sodium, creatinine, BUN, albumin, and hematocrit and her urinary-specific gravity, osmolality, and sodium are elevated. These results suggest that she has hypernatremic dehydration.

6 *Why didn't Mrs. D. complain of thirst?*

The osmoreceptors in the thirst center of the hypothalamus lose their sensitivity with aging. Older adults may become severely dehydrated without feeling a sense of thirst. Further, Mrs. D. was kept on Percocet and Xanax during the week. These medications depress the central nervous system, and they kept Mrs. D. too drowsy to recognize her thirst. Her hyponatremia further contributed to her lethargy and confusion.

7 *How will Mrs. D. be treated?*

Slow rehydration over 48 hours is needed. If Mrs. D. were rapidly rehydrated, the increased osmotic activity in brain cells (from dehydration) could draw in a large amount of water, which could cause cellular swelling and rupture and result in cerebral edema. She is given an IV of 5% dextrose in 0.9% NaCl. Her serum glucose and calcium levels will be monitored carefully as hyperglycemia and hypocalcemia can occur with this method.

Mrs. D. is kept on a careful watch the following week. John observes that the woman who does the dusting takes the water mug and waters the flowers that Mrs. D.'s daughter brought in on Sunday. John asks, "By the way, do you usually take care of the flowers for Mrs. D.?"

"Oh yes!" replies the young woman. "Mrs. D. get flowers *every* Sunday. It's so dry in here that I need to make sure they get water. So I take Mrs. D.'s leftovers in her mug and water the flowers about three times a day. That way, they stay beautiful until Friday afternoon when I throw them out."

"Aha," said John to himself, "a case of unintentional institutional dehydration! Her dehydration during the week contributed to her confusion. The staff during the week assumed that she was incontinent because she was confused. They left her in the wheelchair, which caused her pain. The Percocet added to her confusion and lethargy, as did the Xanax. The lactulose reduced her water absorption, and her water intake was reduced because a staff person diverted her drinking water to the flowers. On the weekends the (different) staff members took their time with her and took her to the toilet every 2 hours. She did not need pain medication and she was not anxious. She drank more fluids in part because her family was there and in part because no one took her water away for the flowers."

References

Ellsbury, D.L., & George, C.S. (2006). Dehydration. *eMedicine,* March 30. http://www.eMedicine.com/ped/topic556.htm

Wixted, T. (2005). A theory about why we forget what we once knew. *Current Directions in Psychological Science, 14*(1), 6–9.

The Trainer

Margie O. is "fifty, fit, and fabulous." She celebrated her big "5-0" day by going to the health club and having a free "quick cholesterol check." Her trainer reported that her cholesterol was 222. "This is too high for someone as fit as you. I suggest you do double sets of your regimen for a month. Sign up for an extra kick boxing class. That will really burn calories and probably lower your cholesterol!" Margie tried this, as well as eating a vegetarian low-fat diet.

The next month she asked her trainer to test her cholesterol again. "Wow—now it's up to 230, Margie! I think you should go to your doctor and have it tested."

Margie had a physical exam, mammogram, colonoscopy, and fasting blood test the following week. The results of her physical exam, mammogram, and colonoscopy were negative. Selected results of her fasting blood test are given below.

Test	Value	Reference Range	Units
Glucose	78	70–100	mg/dL
BUN	8	7–18	mg/dL
Creatinine, serum	1.3	0.8–1.3	mg/dL
Bilirubin, total	1.0	0.1–1.0	mg/dL
Protein, total	7.5	6.4–8.2	gm/dL
Albumin	4.4	3.4–5.0	gm/dL
Alkaline phosphatase	123	50–136	U/L
AST/SGOT	27	15–37	U/L
ALT/SGOT	40	30–65	U/L
GGT	39	1–94	U/L
Cholesterol, total	234	50–199	mg/dL
Triglycerides	150	15–149	mg/dL
LDL (direct)	170	0–99.0	mg/dL
HDL	60	> 40	mg/dL
WBC	7.1	4.5–11.0	K/cmm
RBC	5.0	4.00–5.20	K/cmm
HGB	14.9	12.0–16.0	g/dL
HCT	44.8	36.0–46.0	%
MCV	90	80–100	fL
MCH	32.1	26.0–34.0	pg
MCHC	31.2	31.0–37.0	g/dL
Sodium	140	136–145	mmol/L
Potassium	4.6	3.5–5.1	mmol/L
Chloride	99	98–107	mmol/L
CO_2, total	26.9	21.0–32.0	mmol/L
T_4, free	0.6	0.8–1.5	ng/dL
TSH	6.7	0.4-4.2	μIU/L
Thyroid peroxidase antibody (TPO antibody)	1089	<35	U/mL

Questions

Name ______________________ Section ________

1 Which of Margie's lab values are outside the normal range?

2 What problem does Margie likely have?

3 What is the cause of Margie's problem?

4 What are the typical symptoms of Margie's problem?

5 How will Margie's problem be treated?

6 What other disease can result from too-aggressive treatment of Margie's problem?

Answers to Questions

1 *Which of Margie's lab values are outside the normal range?*

Margie's total cholesterol and low density lipoprotein (LDL) cholesterol are high. Her high density lipoprotein (HDL) cholesterol is in the optimal range. Margie's free T_4 is low, and her thyroid stimulating hormone (TSH) is high. She has elevated thyroid peroxidase antibodies.

2 *What problem does Margie likely have?*

Margie's lab values suggest that she has hypothyroidism with hyperlipidemia.

3 *What is the cause of Margie's problem?*

Hypothyroidism is the most common chronic condition among women aged 50 and older. Approximately 5% of adults in the United States have hypothyroidism. Margie has primary hypothyroidism, an autoimmune disease in which elevated thyroid peroxidase antibodies attack thyroid peroxidase (TPO). TPO is an enzyme in thyroid follicle cells that iodinates T_4 and T_3. As this enzyme is destroyed, inflammation results, which in turn damages the thyroid gland.

Another term for this type of hypothyroidism is Hashimoto's disease, or Hashimoto's thyroiditis. The cause is unknown, but it is 5–10 times more common in women than men and may be related to both a genetic predisposition and declining estrogen with menopause. The disease is "clustered" in families that have other autoimmune diseases such as Type 1 diabetes and celiac disease.

Location of the Thyroid Gland

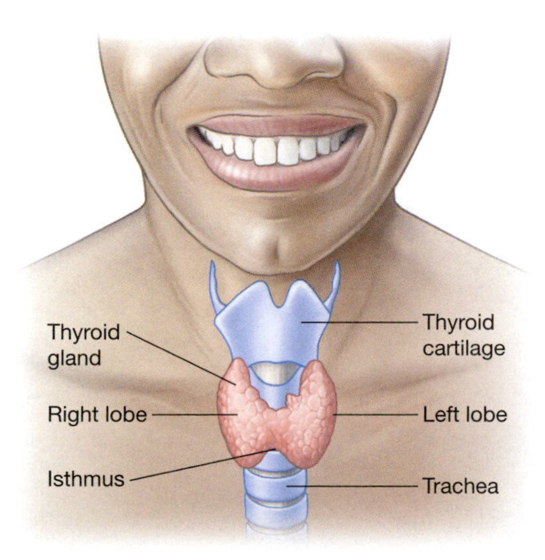

For reasons not yet understood, hypothyroidism is associated with elevated LDL cholesterol. All individuals with elevated LDL cholesterol should have their T_4 and TSH levels measured to rule out hypothyroidism as a cause of hyperlipidemia.

Secondary hypothyroidism is caused by insufficient release of thyroid stimulating hormone (TSH) from the pituitary gland. Pituitary tumors are the most common cause of secondary hypothyroidism. Iron overload (e.g., hemochromatosis) and certain inflammatory diseases can also damage the pituitary and lead to secondary hypothyroidism.

Hypothalamic – Pituitary – Thyroid Axis

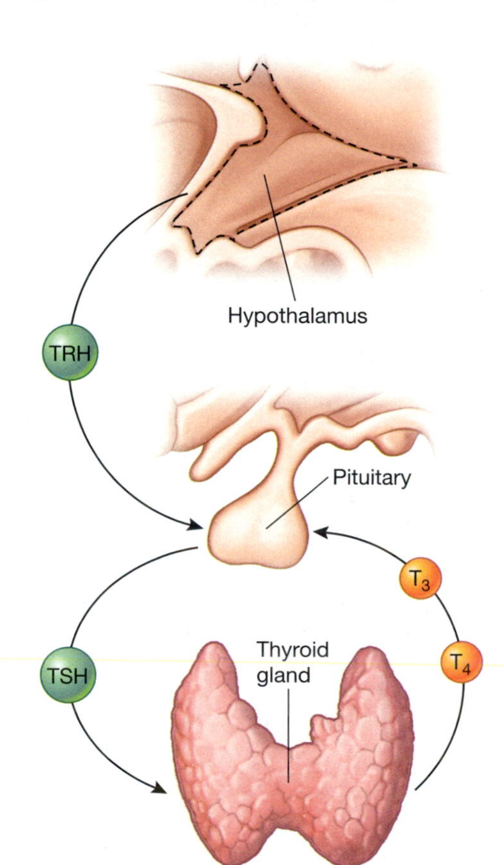

4 *What are the typical symptoms of Margie's problem?*

Margie is a highly fit and energetic woman. She did not have an enlarged thyroid (goiter) or notice any of the following common symptoms of hypothyroidism:

- fatigue
- intolerance to cold
- dry skin, dry scalp
- thinning hair or eyebrows
- facial edema
- weight gain
- heavy menstrual flow or mid-cycle bleeding
- constipation

- husky voice
- muscle cramps

Margie did say that she has "trouble concentrating" at work. She thinks her face is puffier than it used to be—changes she chalked up to menopausal symptoms. She also is having "hot flashes" with menopause, which fluctuate from being too hot to too cold if they occur at night.

5 *How will Margie's problem be treated?*

Margie will be prescribed a thyroid replacement hormone such as levothyroxine. She will be told to take it the first thing in the morning on an empty stomach with a full glass of water. She will take a low dose at first, and the dose will be titrated upward slowly at 6-week increments until her T_4 and TSH levels reach the normal range.

6 *What other disease can result from too-aggressive treatment of Margie's problem?*

Treatment of hypothyroidism increases the risk of osteoporosis. Excess thyroid hormone causes osteoclastic bone resorption and decreases intestinal calcium absorption. Care will be taken to avoid too-low TSH from levothyroxine treatment because over-suppression of TSH greatly increases the risk for osteoporosis. Margie will be advised to eat a diet that is high in calcium, along with a calcium/vitamin D supplement. She must remember to separate her calcium supplement and levothyroxine by at least 4 hours because divalent minerals such as calcium, iron, and magnesium impair the absorption of levothyroxine.

After 3 months of levothyroxine treatment, Margie's T_4 and TSH are in the normal range and her LDL cholesterol came down to 111 mg/dL. Her primary care provider tells her that this fitness regimen and her high HDL greatly reduce her cardiovascular risk. Because her triglyceride level did not change, Margie is asked about her alcohol and sugar intake.

"I love my Margaritas! I guess I have about two a day after working out. But alcohol is good for the heart, right?"

She is told that one drink per day five days of the week (1 oz of hard liquor, or 4 oz of wine) will raise HDL, but drinking more than that actually increases triglycerides. A high sugar intake also can increase triglycerides. Margie is advised to cut back to no more than five drinks per week and to choose sugar-free Margarita mix or switch to wine.

"I can easily do that! I thought alcohol was like exercise—if some is good, more is better!"

References

Thyroid.org
http://www.thyroid.org/patients/brochures.html

National Cholesterol Education Program
http://www.nhlbi.nih.gov/chd/

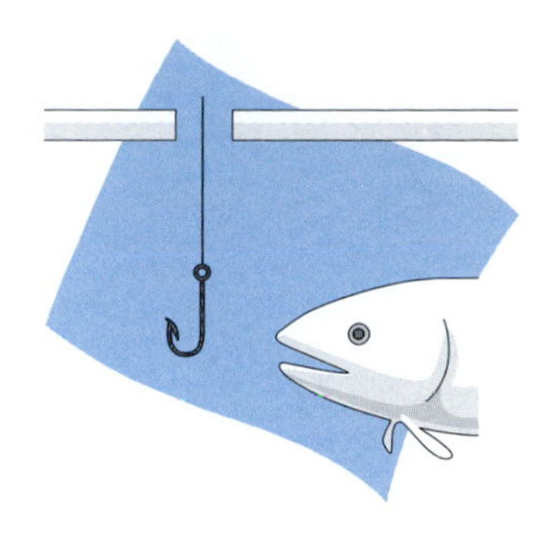

14

The Ice Fishing Derby

It was the day of the Crystal Lake Ice Fishing Derby. Pierre Larouc had been on the ice for 6 hours. He had just finished the traditional derby lunch of hot coffee, baked beans, hot dogs, and s'mores cooked over a fire on the ice. After having two beers and sitting in his ice fishing shanty with a cigar to relax, flags went up simultaneously on three lines. As he ran to check his catch, he felt a crushing pain just below his sternum and radiating to the left shoulder and jaw.

Pierre took a small canister from his pocket and sprayed it into his mouth. Just as fast as the pain came on, it stopped and he pulled up three of the largest pike he had ever caught in February. When he got home to show his catch to his wife Suzy, he found a note from a neighbor stating that she had taken Suzy to the emergency department (ED) because she was dizzy and "lightheaded."

Questions

Name ______________________________ Section __________

1 What did Pierre spray into his mouth? What did it do?

2 What pathophysiological problem does Pierre have? What causes it?

3 Did Pierre have a myocardial infarction (MI, or heart attack)? What leads you to this conclusion?

4 What triggered the pain that Pierre experienced? Explain the mechanism.

5 When Suzy was seen in the ED, her blood pressure was 90/58. Her normal blood pressure is 115/70. What do you think caused this episode of hypotension?

6 What precautions must Pierre remember to take in the future?

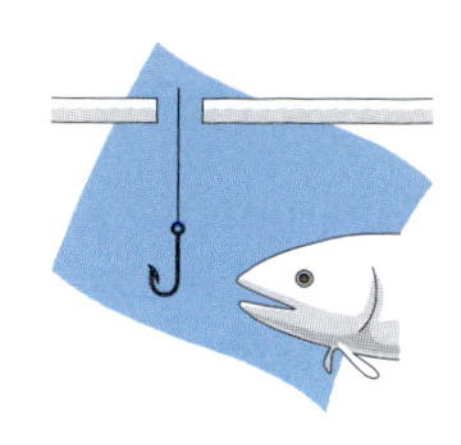

Answers to Questions

1 *What did Pierre spray into his mouth? What did it do?*

Pierre sprayed nitroglycerin under his tongue. The nitroglycerin was absorbed directly into the systemic circulation through the lingual mucosa. (Nitroglycerin can be administered by buccal, sublingual, inhalation, and transdermal routes, but it is ineffective if swallowed. It is destroyed by gastric acid, and what little is absorbed is metabolized by the liver before it gets to the systemic circulation.) Nitroglycerin increases coronary blood flow by dilating coronary arteries. It also produces systemic vasodilation (especially venous).

Nitroglycerin Spray

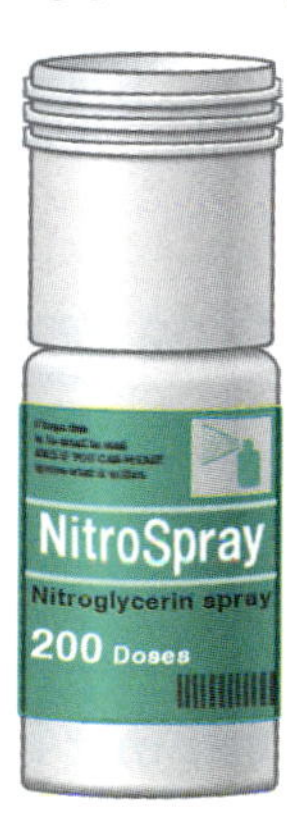

2 *What pathophysiological problem does Pierre have? What causes it?*

Pierre has "classic" angina. Classic angina is basically a condition of diminished oxygen supply and/or increased oxygen demands in the myocardium. It results from atherosclerosis in the coronary arteries (see illustration on page 124). Angina also can result from vasospasm in one or more coronary arteries. This type of angina, called "variant" angina, typically occurs during rapid eye movement (REM) sleep in the early morning. Anginal pain also can result from increased metabolic demands such as during thyrotoxicosis.

Some people with valvular problems such as aortic stenosis or mitral regurgitation develop angina. Aortic stenosis (narrowing of the aortic valve) blocks blood flow from the left ventricle. This reduces perfusion pressure in the heart and causes anginal pain. People with mitral regurgitation (leaking mitral valve) can develop anginal pain from the same triggers that cause classic angina.

Angina results in myocardial ischemia, causing a cascade of events including release of potassium, histamine, serotonin, and lactic acid from the myocardial cell. All of these in turn activate nerve endings and cause pain. Norepinephrine released in response to the pain increases thromboxane A2 release which stimulates platelet aggregation and increases the risk of thrombosis.

Below are his lab values.

Lab Test	Mr. Lee's Values	Normal Values
Calcium	11.2 mg/dL	8.5–10.5 mg/dL
Phosphorus	2.8 mg/dL	2.0–5.0 mg/dL
Potassium	5.3 mEq/L	3.2–5.2 mEq/L
Sodium	139 mEq/L	135–145 mEq/L
Alkaline phosphatase	145 U/L	25–125 U/L
Alkaline phosphatase (bone fraction)	90 U/L	11–73 U/L
Alkaline phosphatase (liver fraction)	50 U/L	0–93 U/L
Thyroxine (T_4) total	10 ug/dL	4.6–10.5 μg/dL
TSH	2.5 uIU/L	0.4–4.2 μIU/L
Parathyroid hormone (intact)	155 pg/mL	10–65 pg/mL

Gerome is sent for a Sestamibi scan of the neck. Sestamibi is a small protein radio-labeled with technetium-99, injected into the vein. Technetium-99 is taken up by the parathyroid if there is a high concentration of parathyroid hormone, as in parathyroid tumors. The radioactivity can be detected and visualized by a gamma ray detecting scanner.

Following is an image from Gerome's scan.

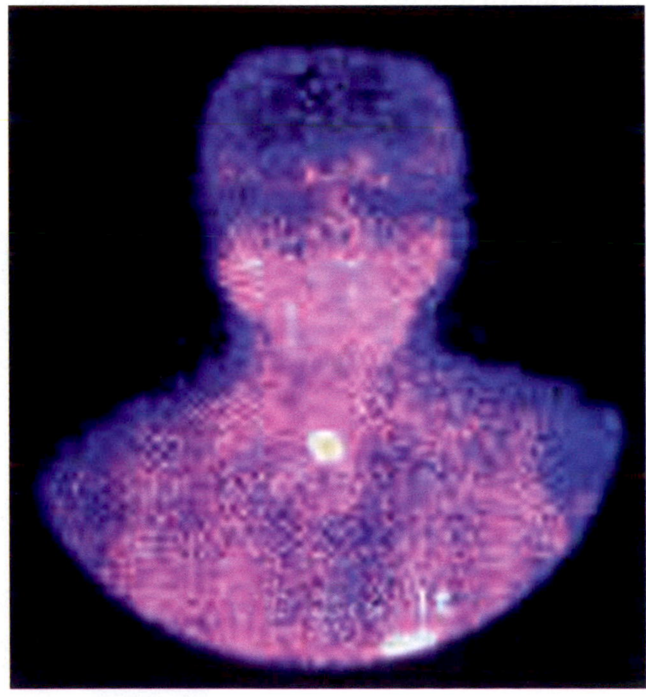

Courtesy of J. Norman, MD; www.parathyroid.com

Questions

Name ______________________________ Section __________

1 What is the difference between osteopenia and osteoporosis, and how can each be diagnosed?

2 What are the risk factors for bone loss in a 50-year-old man?

3 What do Gerome's lab values suggest? What symptoms support this conclusion?

4 What does the Sestamibi scan of the neck region show?

5 How will Gerome be treated?

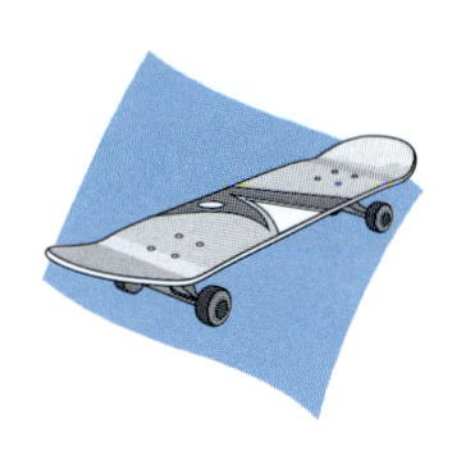

Answers to Questions

1 *What is the difference between osteopenia and osteoporosis, and how can each be diagnosed?*

Osteopenia and osteoporosis represent different degrees of bone loss. A bone mineral test using dual-energy X-ray absorptiometry, the DEXA test, can quantify bone density. The so-called T-score is based on standard deviations from normal bone mass of a 30-year-old healthy adult. A T-score of –1.0 and greater (1 standard deviation and less) is set as "normal" bone mineral density. Osteopenia is defined as having between –1 and –2.5 standard deviations below normal bone mass. Osteoporosis is defined as –2.5 standard deviations below normal. Gerome's DEXA score was found to be –2.0 in the bones of his wrist and forearm and in the neck of his femur bone and –1.5 in his lumbar vertebrae, indicating osteopenia.

Sites for Bone Density Measurements

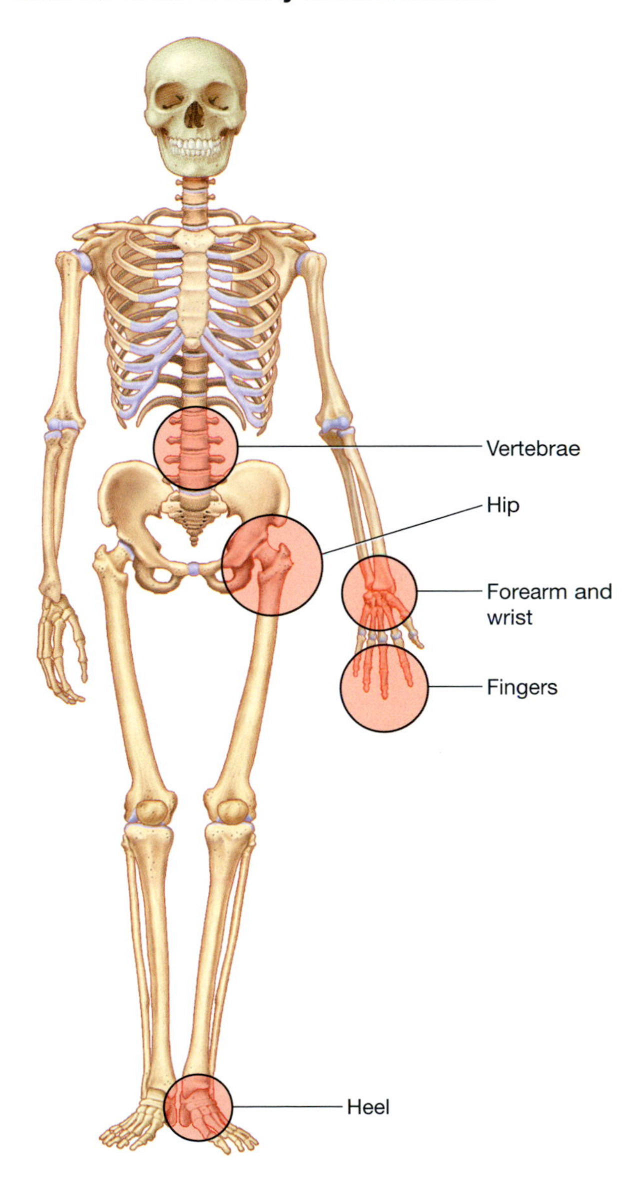

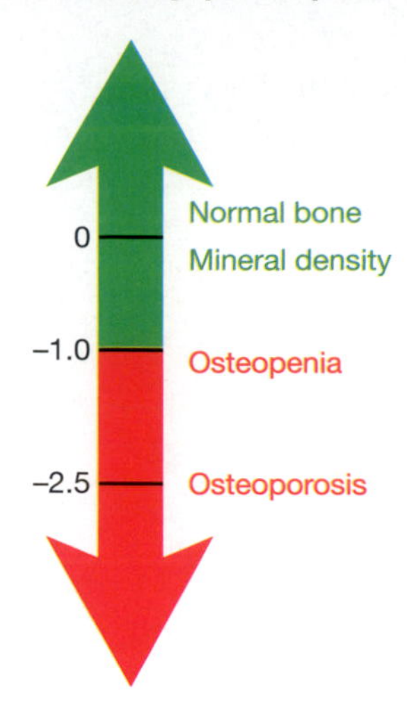

2 *What are the risk factors for bone loss in a 50-year-old man?*

Bone loss is not seen only in post-menopausal women. Approximately 25% of men aged 50 and older eventually will incur a bone fracture caused by osteoporosis. Osteoporosis is preceded by osteopenia—often for decades. Risk factors for bone loss in a man include:

- Caucasian or Asian race
- small, thin body
- inactive lifestyle
- diet low in dairy products
- long-term use of glucocorticosteroids, such as prednisone
- long-term use of aluminum-containing antacids
- cigarette smoking
- hyperthyroidism or hyperparathyroidism

Gerome is Asian. He has a thin body and an inactive lifestyle as a computer programmer. His only exercise is walking the dog when Josh is not available. He is lactose-intolerant and does not consume dairy products other than Parmesan cheese. He does not take a calcium supplement. He took oral prednisone for asthma for 4+ years between the ages of 8 and 12. Before his peptic ulcer was diagnosed, he self-medicated with aluminum-containing antacids for at least 3 years.

3 *What do Gerome's lab values suggest? What symptoms support this conclusion?*

Gerome has an elevated serum calcium and alkaline phosphatase (bone fraction). His parathyroid hormone is greatly elevated. Taken together, these lab values suggest that he has hyperparathyroidism. Subperiosteal resorption of cortical bone, especially at the radial aspect of the middle phalanx of the index finger and the middle finger, is characteristic of hyperparathyroidism.

"Brown tumors" are rare characteristics (fewer than 2% of cases). Gerome has a brown tumor in the middle phalanx of his left index finger. Brown tumors are benign; they develop in areas where the osteoclasts are overactive. They indicate an area where the bone has been replaced with fibrous tissue, bone, and blood vessels but no bone matrix and show up as brown spots ("radiolucent") on X-ray.

Other symptoms of hyperparathyroidism are kidney stones (Gerome has had two episodes in the past 6 months) and peptic ulcer (he was treated for peptic ulcer 2 years ago and still complains of gastric pain.) The mnemonic "painful bones, renal stones, abdominal groans, and psychic moans" help one to remember the symptoms of hyperparathyroidism. Many people remain undiagnosed until an X-ray is taken following an injury.

4 *What does the Sestamibi scan of the neck region show?*

Radio-labeled Sestamibi is taken up by overactive parathyroid glands. (*Note:* This radio labeled protein is the same one used in cardiac stress tests. It does *not* cross-react with other X-ray dyes and so is safe for patients with previous allergic reactions to dyes.) In the presence of high serum calcium levels, normal parathyroid glands undergo negative feedback and become inactive; they do not take up the radio-labeled protein. Gerome has a parathyroid tumor. About 80% of parathyroid tumors are "benign" adenomas—benign in the sense of not being cancerous—but they are dangerous and must be removed surgically to prevent severe osteoporosis and calcium deposits in soft tissue such as the kidney and heart.

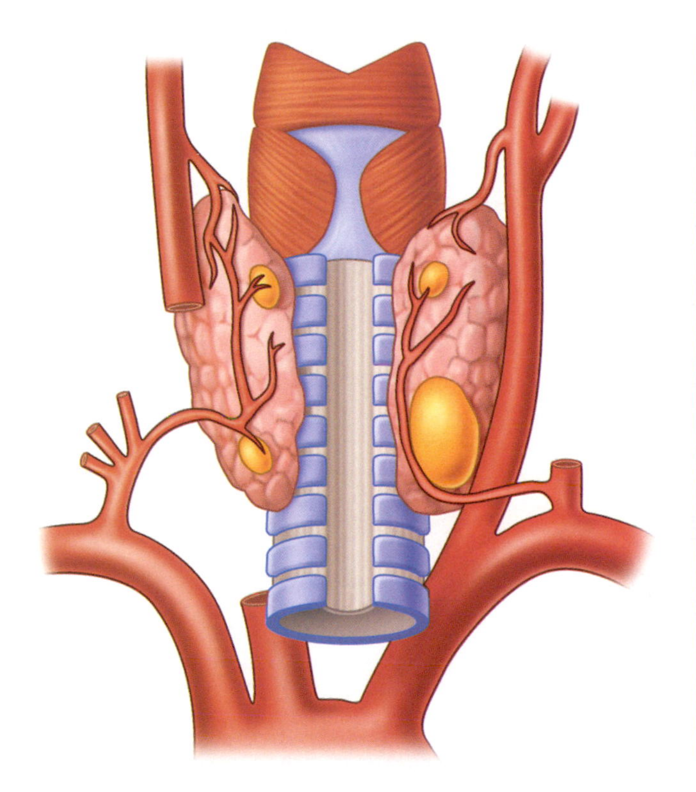

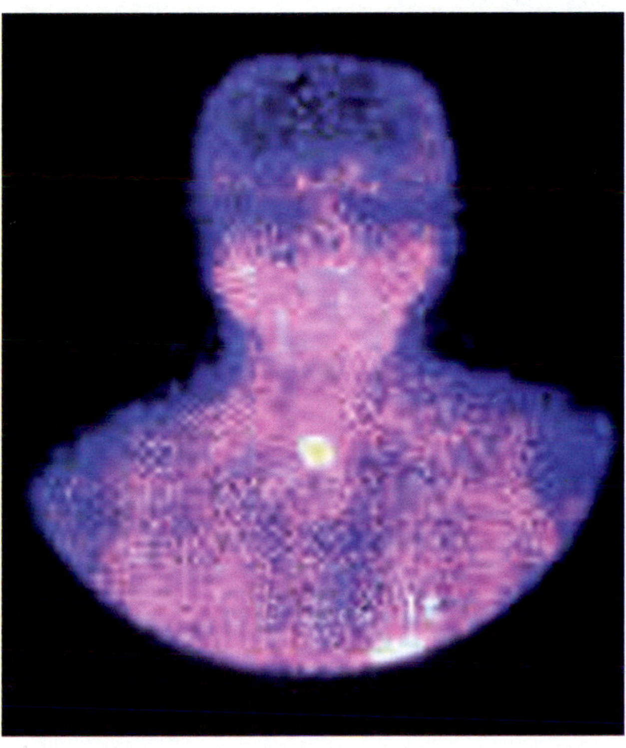

Courtesy of J. Norman, MD; www.parathyroid.com

Gerome's Parathyroid Adenoma

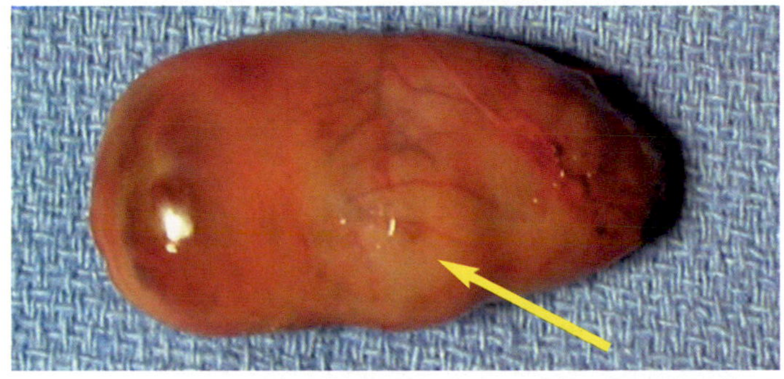

Courtesy of J. Norman, MD; www.parathyroid.com

5 *How will Gerome be treated?*

Gerome underwent minimally invasive parathyroid surgery. Removal of a benign adenoma was confirmed by the pathologist. One month after the parathyroid adenomectomy, Gerome's calcium, alkaline phosphatase and parathyroid hormone levels were in the normal range. He quit smoking and walks the dog every day. He is doing weight training under the supervision of a physical therapist. He takes a calcium supplement with vitamin D and vitamin K twice a day. (Vitamin K now is known to be essential to develop and maintain bone mass. The supplement is best taken with a meal, as fat improves absorption of fat-soluble vitamins such as D and K.) "I've really done a 180 on my habits!" he exclaimed.

References

Osteopenia. WebMD
http://www.webmd.com/osteoporosis/tc/Osteopenia-Overview

Salen, P.M. (2006). Hyperparathyroidism.
http://www.eMedicine.com/emerg/topic265.htm

Sestamibi Scans
http://www.parathyroid.com/sestamibi.htm

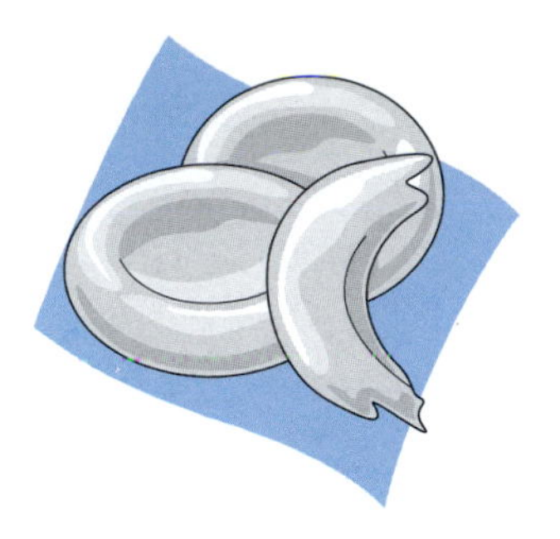

16

The Uninformed Coach

Shawn has sickle cell disease (SCD), also known as sickle cell anemia. He grew up knowing how serious SCD can be, but he was fortunate. He was a healthy child with no infections other than the occasional cold. His parents and brother and sister doted on him as the youngest and smallest child in the family. In high school he tried out for the track team and was able to compete in the shot put. He wasn't "the best," but he was the only person who did the shot put on his team. When he went to college, this was the first time he had been away from home other than his yearly week away at the Hole in the Wall Gang Camp.

In college, Shawn liked his classes, team, and coach. It was great to be on his own! In mid-October he felt the cold while walking across campus. At first it felt good, with the fall leaves against the blue sky. By November, when he caught a common cold, he developed a pain in his left thigh. At the indoor practice the coach told him to work through it and then put a heat pack on it: "You've probably pulled a muscle from throwing the way you do. We're going to help you unlearn your bad habits. By spring you'll be able to throw twice as far."

When Shawn got back to his room, his left leg hurt more. He called his mom, who drove to the campus and took him to the ED. By the time he was seen, he was in so much pain that he couldn't speak. His upper abdomen hurt as much as his leg. He was having difficulty breathing.

Shown here is a photomicrograph of Shawn's blood smear.

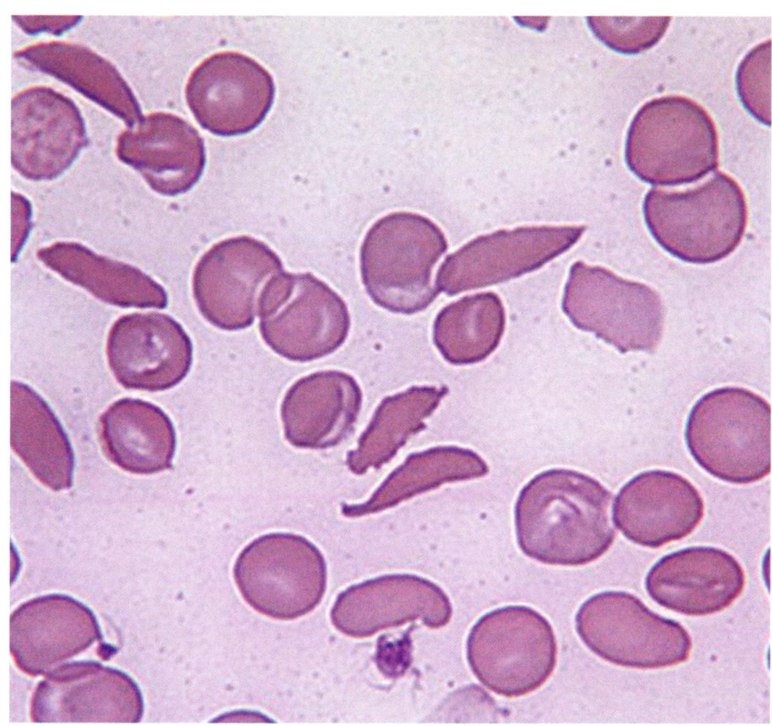

Shawn's X-ray is below.

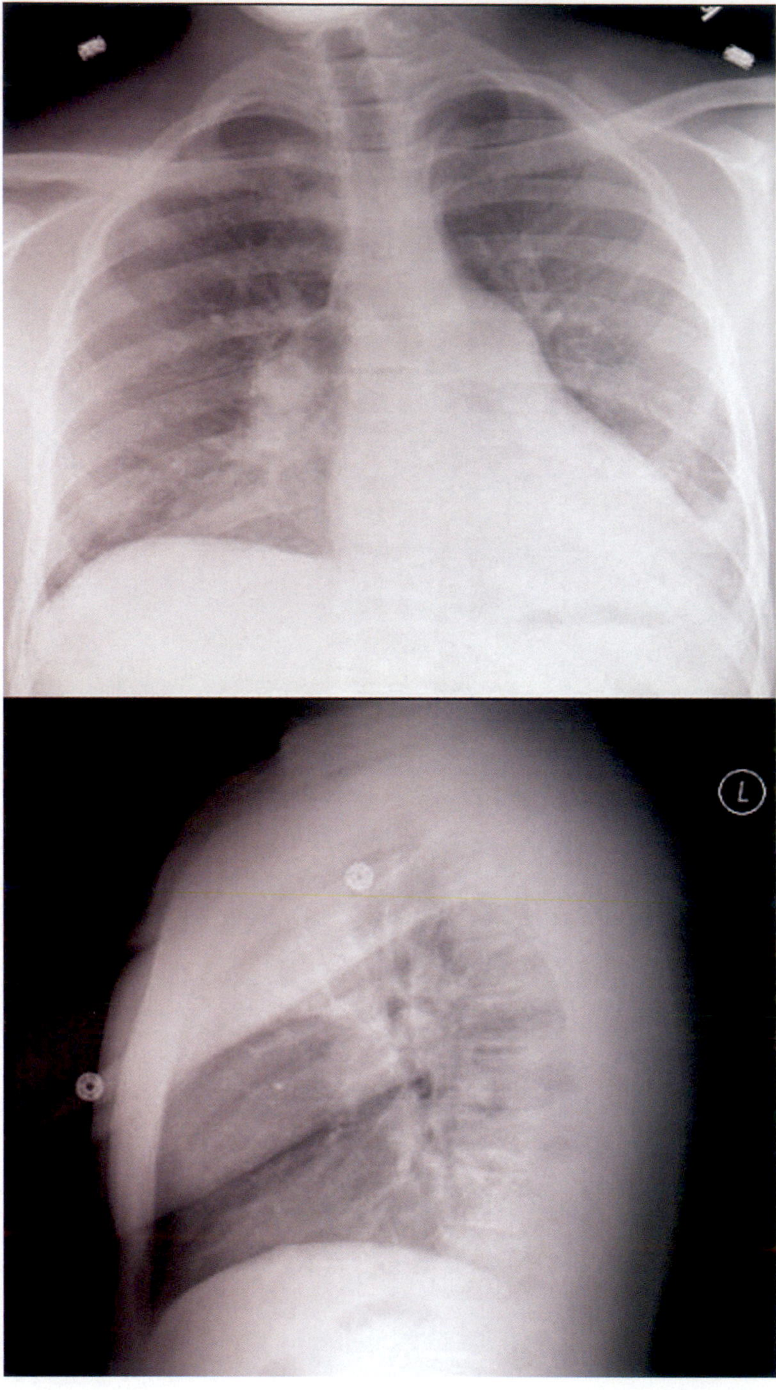

Courtesy of www.pediatriceducation.org

Questions

Name ______________________________ Section __________

1 What is the Hole in the Wall Gang Camp?

See: www.holeinthewallgang.org

2 How is sickle cell disease contracted?

3 What are the possible complications of sickle cell disease?

4 What acute complication of sickle cell disease does Shawn probably have?

5 How will Shawn be treated for this complication?

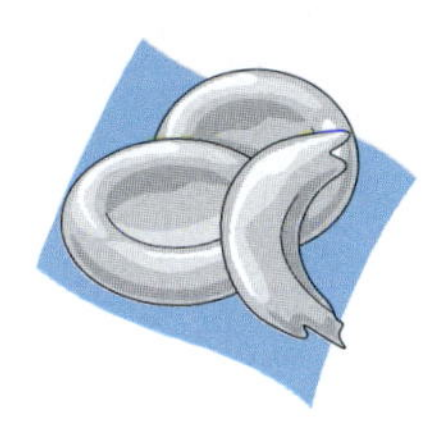

Answers to Questions

1 *What is the Hole in the Wall Gang Camp?*

See: www.holeinthewallgang.org

The Hole in the Wall Gang Camp (named after Butch Cassidy's fabled camp in the classic movie *Butch Cassidy and the Sundance Kid*) was started by the late actor Paul Newman. The camp offers a summer camp experience for children with serious illnesses. Two of the week-long camp sessions are reserved for children with sickle cell disease.

2 *How is sickle cell disease contracted?*

Sickle cell disease is an autosomal recessive disorder. Approximately 8% (1/12) of African Americans carry the trait, but only 1 in 400 have the disease. It affects primarily people of African descent but also occurs in people whose heritage is from the Mediterranean, Middle East, and India. A shift in one nucleotide of the 438 bases that code for the hemoglobin beta chain causes a switch in the sixth amino acid in the chain, resulting in a defective hemoglobin known as hemoglobin S. Deoxygenated hemoglobin S acts like a gel; it stacks into filaments and clusters inside the RBCs, causing the characteristic sickle shape. Sickled RBCs can resume a normal shape when exposed to oxygen, but over time they permanently become distorted.

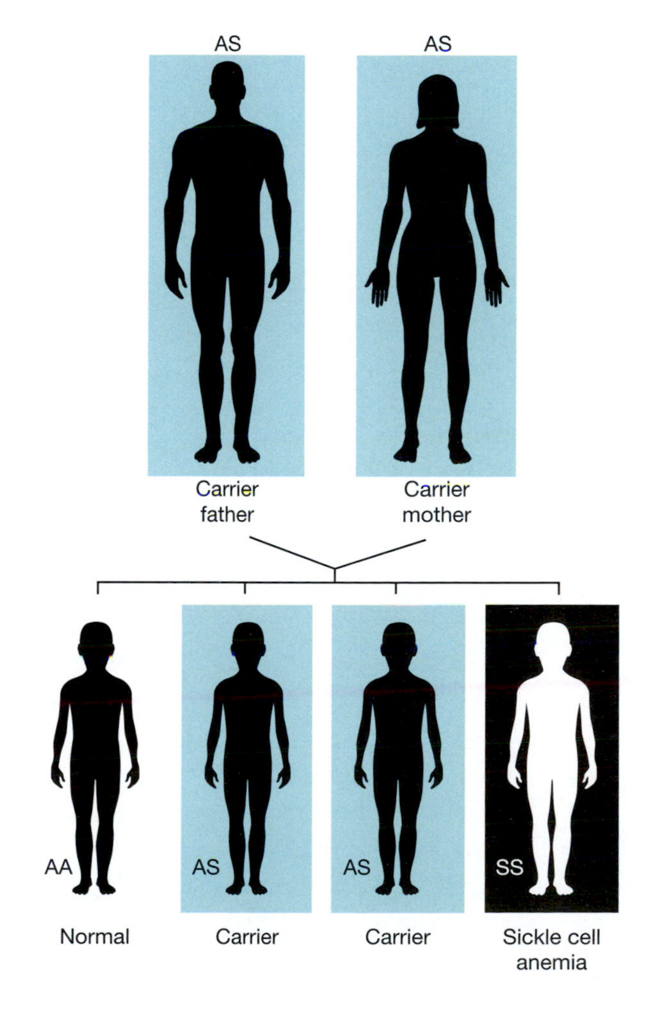

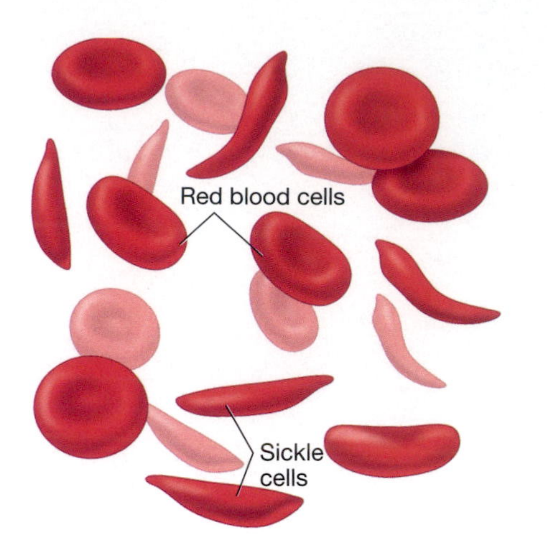

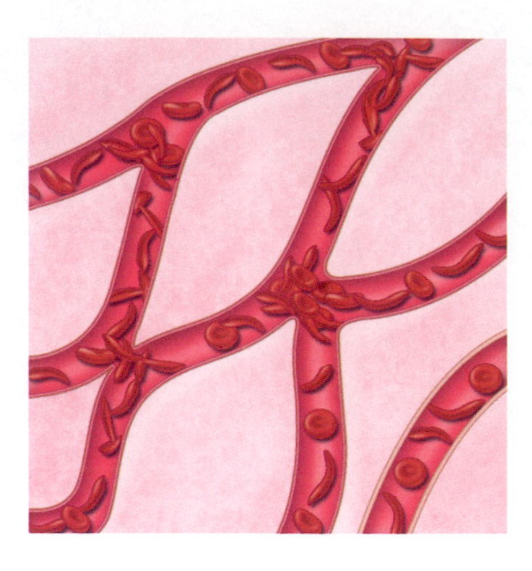

3 *What are the possible complications of sickle cell disease?*

Sickled cells are less efficient in oxygen delivery and they are more fragile. The lifespan of a sickle RBC cell is much shorter than normal, so individuals with sickle cell disease typically have reduced RBC counts (anemia). The sickled cells can collect and obstruct blood vessels, and they increase the viscosity of the blood, which slows blood flow, allowing the formation of thrombi.

The mnemonic HBSS PAIN CRISIS can help explain the common complications:

- H **Hemolysis.** Sickled cells are more fragile than normal RBCs'; they rupture (hemolyze) easily.
- B **Bone marrow hyperplasia.** Bone marrow tries to counter the loss of sickled cells from hemolysis by undergoing hyperplasia.
- S **Stroke.** Sickled cells can collect in small vessels in the brain and cause a stroke.
- S **Skin ulcers.** Blood vessels in the leg that become obstructed by sickled cells can reduce oxygen delivery to skin. Serious ulcers can result.

- P **Pain crises.** Periodic severe pain is caused by obstructions in small blood vessels in the muscle, bone, abdomen, and chest.
- A **Anemia/aplastic crisis.** Aplastic crisis occurs when the bone marrow is unable to replace enough RBCs that were lost to hemolysis. The result is a rapid reduction in RBC count.
- I **Infections.** Common sites of infection are the lungs, bladder, bone/joints, and central nervous system (CNS).
- N **Nocturia.** Urinary frequency during the night is common.

- C **Chest syndrome.** Acute chest syndrome (chest crisis) is characterized by fever, chest pain, tachypnea, hypoxia, and pulmonary infiltrates. It is a leading cause of death in individuals ages 11 and older with sickle cell disease.
- R **Retinopathy/renal failure.** Small blood vessels that supply oxygen to the retina and kidneys can become obstructed.
- I **Infarction.** Obstructed blood vessels (vasoocclusive crisis) can occur anywhere (e.g., bone, spleen, muscle, kidneys, bowel).
- S **Splenic sequestration crisis.** When sickled RBCs hemolyze, the spleen responds by rapidly taking up RBCs ("sequestering"). This causes a rapid decrease in hematocrit, hemoglobin, and blood volume that can result in hypotension and shock. Eventually the spleen destroys itself (autosplenectomy) and the number of events decreases.
- I **Increased risk of fetal loss during pregnancy.** Obstetrical patients are at very high risk. Also, the hydroxyurea used to prevent chest syndrome is teratogenic.
- S **Sepsis.** Patients with bloodborne infections (sepsis) are at high risk for death.

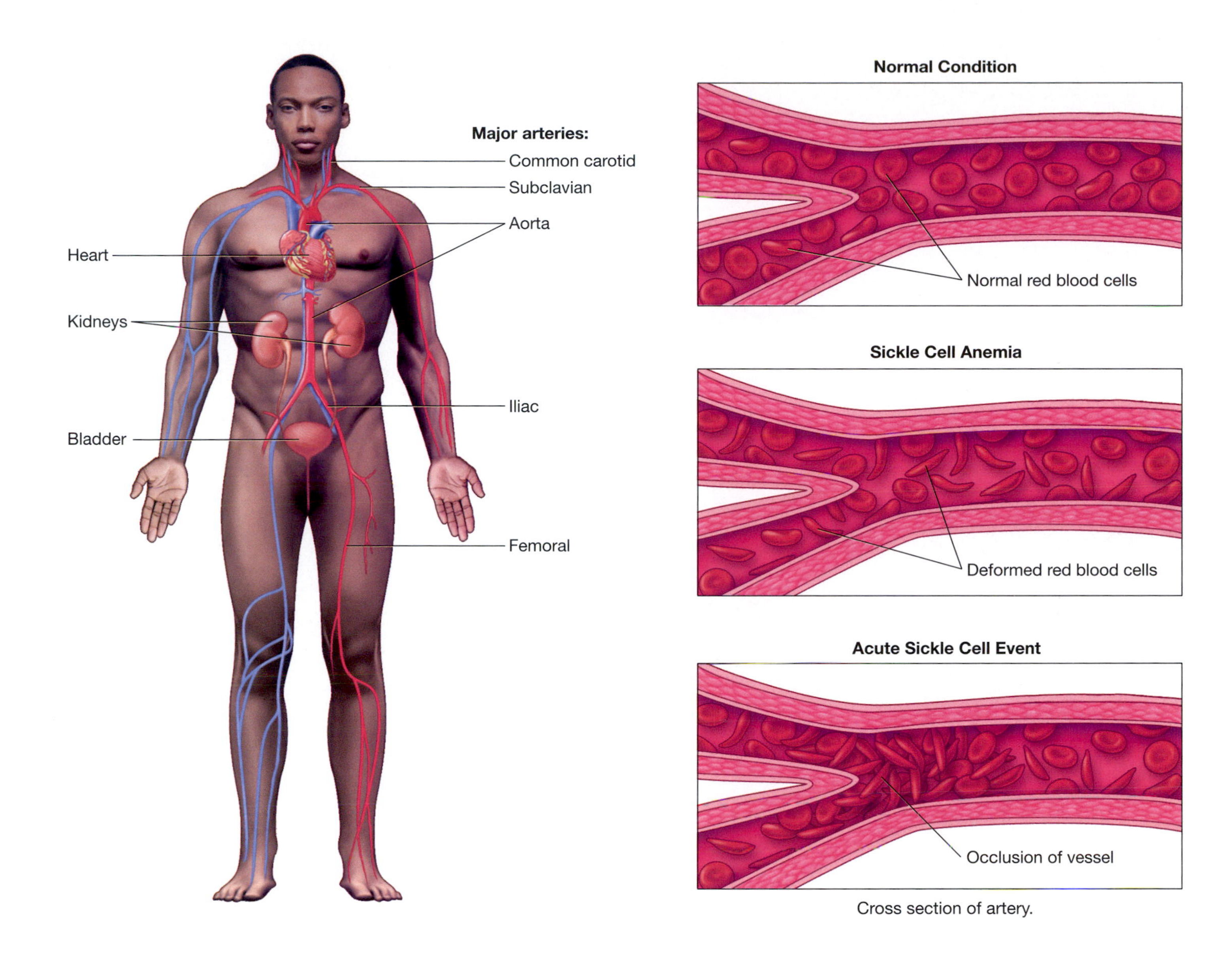

Cross section of artery.

4 *What acute complication of sickle cell disease does Shawn probably have?*

Shawn's symptoms and the infiltrates visible on Shawn's X-ray indicate that he is suffering an acute chest crisis.

5 *How will Shawn be treated for this complication?*

Shawn will be given oxygen therapy to treat hypoxemia. He also will receive a red blood cell transfusion and a normal saline drip. He will be given antibiotics for *Streptococcus pneumoniae*, *Mycoplasma*, and *Chlamydia pneumoniae*. The opioid morphine will be administered for pain control.

He will begin to take a daily supplement of folic acid (folate). This vitamin has been shown to support erythropoiesis (red cell synthesis) in patients with SCD.

Shawn also will start to take hydroxyurea (Droxia) as a prophylactic medication to prevent future acute chest crises. Hyroxyurea stimulates the synthesis of fetal hemoglobin, which is resistant to sickling, thereby reducing the incidence of vasoocclusive crises. The drug reduces the risk of future chest crises by 50% and reduces overall mortality by 40%. In the 1970s the median lifespan of a person born with SCD was only 14 years. Today the median lifespan is 50 years as a result of improved treatment for SCD. Some are concerned; however, that hydroxyurea may increase the risk of leukemia or malignant tumors.

By spring Shawn was able to throw the shot put twice as far. In fact, he helped the team win the regional track title. He had no vasoocclusive recurrences and no further infections. He lined up a summer job at the Hole in the Wall Gang Camp as a camp counselor.

References

American Sickle Cell Anemia Association
http://www.ascaa.org/

Sickle Cell Anemia. MayoClinic.com
http://www.mayoclinic.com/health/sickle-cell-anemia/DS00324/DSECTION=2

Sickle Cell Disease Association of America
4221 Wilshire Blvd.
Los Angeles, CA 90010
1-800-421-8453

Sickle Cell Disease Program Division of Blood Diseases and Resources
National Heart, Lung and Blood Institute
II Rockledge Center
6701 Rockledge Drive MSC 7950
Bethesda, MD 20892-7950
301-435-0055

Hole in the Wall Gang Camp
http://www.holeinthewallgang.org/

Taher, A., & Kazzi, Z.N. (2007). Anemia, Sickle Cell. *eMedicine*, Jan. 11.
http://www.eMedicine.com/emerg/topic26.htm

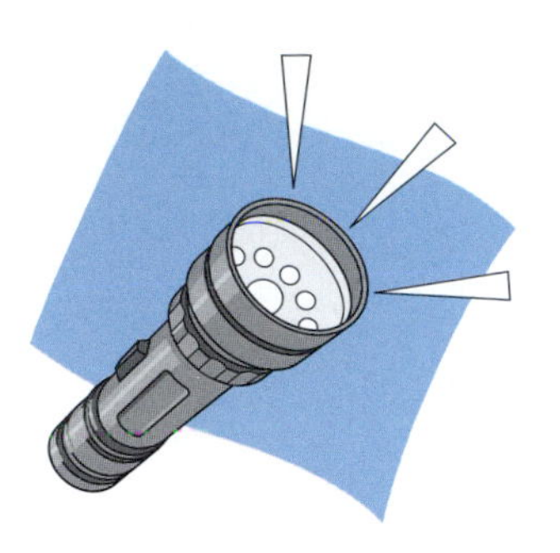

17

Nightwatch

After Mr. D. retired from the police force of a small Midwestern city, he took a job as a night watchman for an insurance company. After 25 years of detective work, rounds on the 30-story building were, in a word, "boring." But he was determined to persevere for 2 years until his wife could retire from her third-shift nursing supervisor's job. Much of Mr. D.'s shift involved watching monitors. His weight had crept upward. "At 275, I'd be in trouble if I were still on the force," he said.

As his weight increased, so did his indigestion. At first antacid tablets seemed to work—but then he realized that he was taking them every 2 hours. He started taking an acid reducer/antacid combination (Pepcid AC), and this seemed to work for a few months. Then he saw an ad on TV for a new over-the-counter acid reducer that promised 'round-the-clock acid reduction. He found a generic version of omeprazole (Prilosec OTC) in the corner store and took that for 3 months.

When he went for his annual physical, he reported "I'm tired all the time. I think it's just my late-shift work—I don't think it suits my biological clock. My other problem is my indigestion. Last Thursday I was eating a corn dog, and the thing seemed to get stuck in my esophagus. Wow, did I suffer that night! I've had hiatal hernia all my life. I keep my head propped up with pillows at night, and I don't go to bed until at least 3 hours after eating. . . . Well, I used to do this anyway, until I took this night job."

Mr. D.'s only prescription drug is amlodipine (Norvasc) 10 mg/day taken for hypertension. His blood pressure was at target at his office visit: 135/80. Mr. D.'s fasting blood values are on the following page. He was also sent to a gastroenterologist, who performed a endoscopic exam. (See Case 18 for a discussion of endoscopic upper GI exams.)

Below are the results of his lab tests.

Test	Value	Reference / Range	Units
Ammonia	47	19–60	mcg/dL
BUN	12	7–18	mg/dL
Creatinine, serum	1.1	0.8–1.3	mg/dL
Bilirubin, total	0.7	0.1–1.0	mg/dL
Bilirubin, direct (conjugated)	0.1	0.1–0.3	mg/dL
Bilirubin, indirect (unconjugated)	0.4	0.2–0.8	mg/dL
Protein, total	7.5	6.4–8.2	gm/dL
Albumin	4.4	3.4–5.0	gm/dL
Alkaline phosphatase	125	50–136	U/L
Aspartate amino-transferase (AST)	29	15–37	U/L
Alanine amino-transferase (ALT)	51	30–65	U/L
Gamma-glutamyl transferase (GGT)	79	1–94	U/L
Cholesterol, total	220	50–199	mg/dL
Triglycerides	149	15–149	mg/dL
LDL, direct	140	0–99	mg/dL
HDL	45	> 40	mg/dL
RBC	4.1	4.50–5.90	K/cmm
Hemoglobin (HGB)	15.5	13.5–17.5	g/dL
Hematocrit (HCT)	53.0	41.0–53.0	%
Mean corpuscular volume (MCV)	105	80–100	fL
Mean corpuscular hemoglobin (MCH)	29	26.0–34.0	pg
Mean corpuscular hemoglobin concentration (MCHC)	32.0	31.0–37.0	g/dL
Prothrombin time (PT)	10	9–12	seconds

Below is a photograph of an abnormal finding on endocsopic exam of the esophagus. A biopsy was taken for further examination.

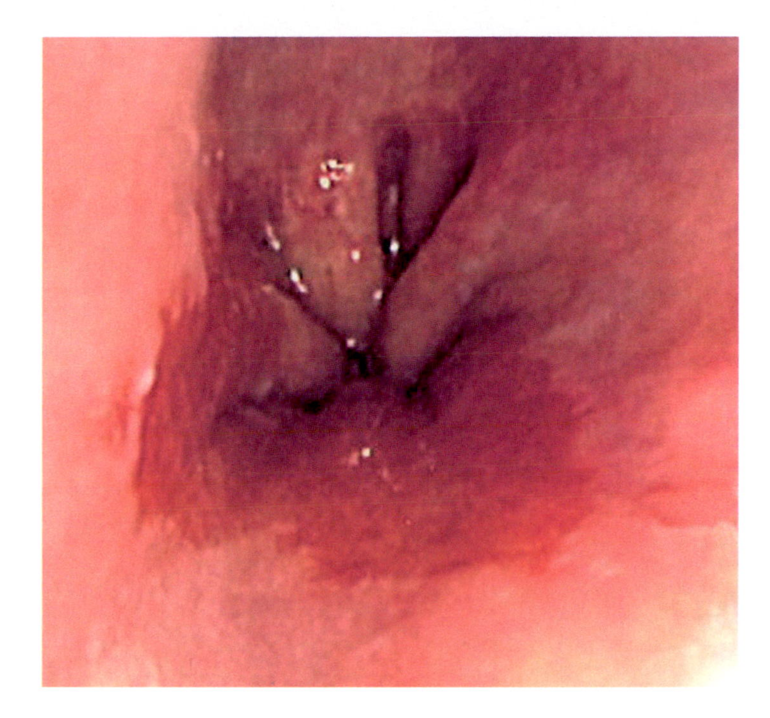

Below is a photomicrograph of the abnormal cells removed from Mr. D.'s esophagus.

Normal

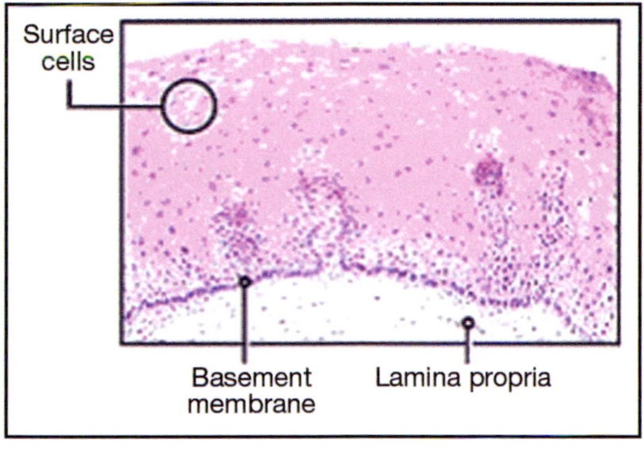

Courtesy of Barretsinfo.com

Mr. D.'s Biopsy

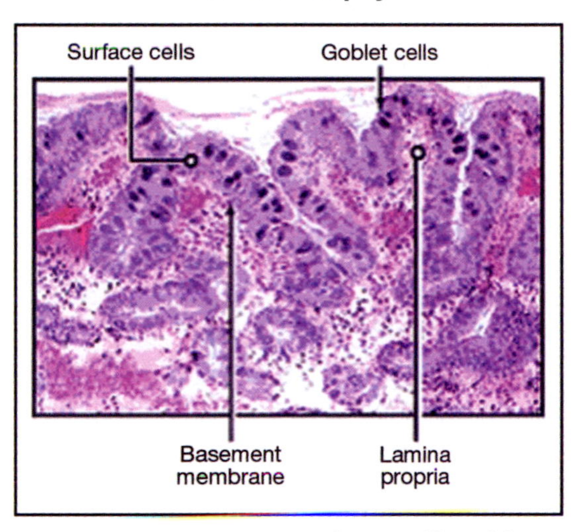

Courtesy of Barretsinfo.com

The biopsy revealed that the normal stratified squamous epithelium was replaced by metaplastic, specialized, columnar glandular epithelium known as "specialized intestinal metaplasia of the esophagus."

Questions

Name ____ Section __________

1 What do Mr. D.'s lab values suggest?

2 What are the possible causes of these abnormal lab values?

3 What other lab values do you think will be ordered to confirm your suspicions?

4 What problems does Mr. D. have with his esophagus? What does the biopsy suggest?

5 What future problem is Mr. D. now at risk for?

6 How will Mr. D. be treated?

7 What lifestyle changes can help to improve Mr. D.'s symptoms?

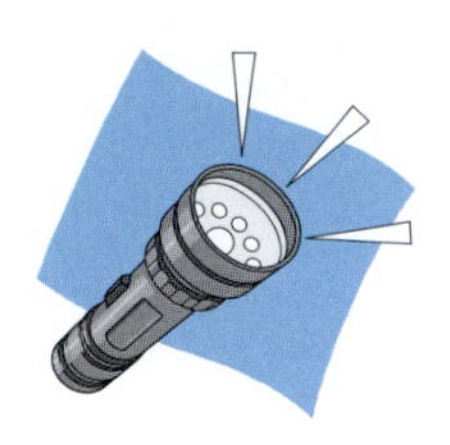

Answers to Questions

1 *What do Mr. D.'s lab values suggest?*

Mr. D. has a high LDL cholesterol. He also has a high mean corpuscular volume (MCV) and low red blood cell (RBC) count indicating macrocytic anemia.

2 *What are the possible causes of these abnormal lab values?*

Mr. D. has begun a night watchman's job where he admits he is bored and eats on the job. He is obese and is eating foods such as corn dogs that are high in saturated fat and calories. Over time, increased weight and consumption of foods high in saturated fat can increase LDL cholesterol.

Macrocytic anemia may result from a folate and/or vitamin B_{12} deficiency. Folate deficiency may be caused by diets low in green leafy vegetables, fortified breads and cereals, or drug-induced folate deficiency. Vitamin B_{12} deficiency is not likely a result of dietary causes, as Mr. D. eats animal proteins with adequate vitamin B_{12} content. Dietary vitamin B_{12} is bound to animal proteins and gastric acid is necessary to cleave the vitamin off the protein so it can bind to intrinsic factor and be ferried to the jejunum for absorption. Mr. D. has been self-medicating with antacids and acid reducers for a long time and may not have sufficient gastric acid to be able to cleave vitamin B_{12} from animal protein foods.

3 *What other lab values do you think will be ordered to confirm your suspicions?*

Serum folate and vitamin B_{12} levels were ordered. Mr. D.'s folate level was normal, but his vitamin B_{12} level was low.

4 *What problems does Mr. D. have with his esophagus? What does the biopsy suggest?*

Mr. D. has Barrett's esophagus. Normally the esophagus is lined by squamous cells. An area in Mr. D.'s esophagus has been replaced by cells typically found in stomach and intestinal epithelium. He has specialized intestinal metaplasia. The term metaplasia refers to tissue that normally is not found in a particular region. In Mr. D.'s case, the esophagus has regions of specialized tissue usually found in the intestine. Note the goblet cells that are stained blue in the micrograph. Actual "glands" of goblet cells can form in the epithelial tissue of the esophagus as Barrett's esophagus progresses.

Mr. D. reports that he has had hiatal hernia for many years. Hiatal hernia is an extension of the stomach above the diaphragm. The lower esophageal sphincter is weaker in people with hiatal hernia. This allows stomach acid and bile to splash up into the esophagus and cause a problem known as gastroesophageal reflux disease (GERD). Most people with symptoms of GERD, commonly called "heartburn," have hiatal hernia, but most people with hiatal hernia do not have GERD. It is thought that, over time, GERD can lead to the cellular changes seen in Barrett's esophagus. The injury/inflammation from GERD causes cells deep in the wall of the esophagus to transform into metaplastic cells that mimic those found in the stomach and small intestine.

The diagram below shows hiatal hernia.

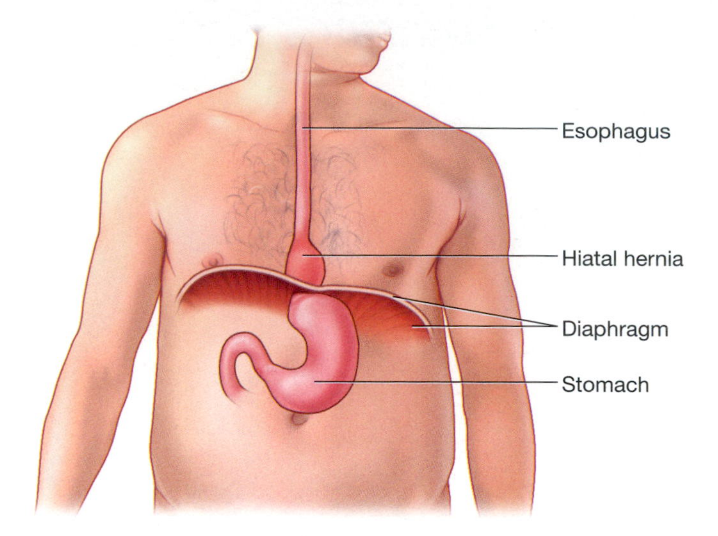

The area of specialized intestinal metaplasia was found above the hiatal hernia.

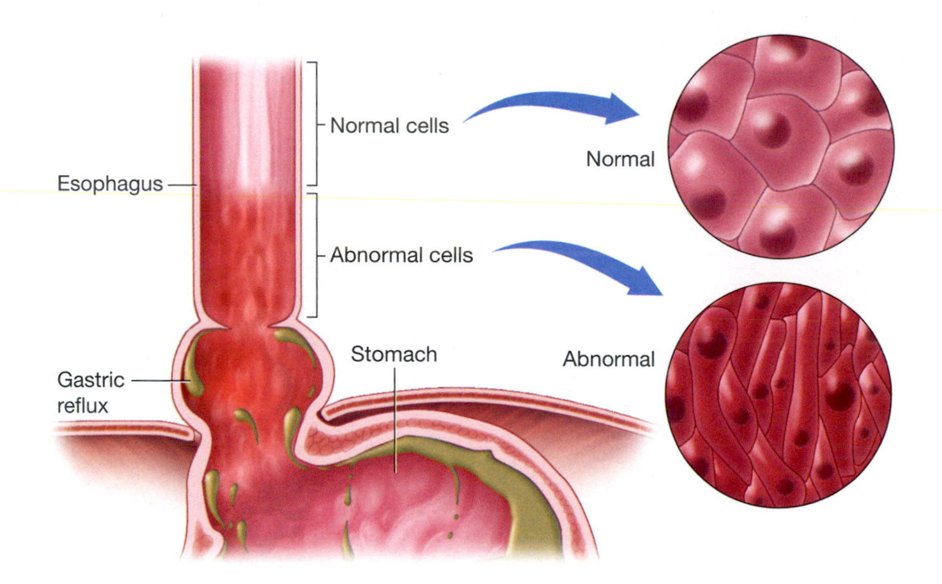

5 *What future problem is Mr. D. now at risk for?*

Barrett's esophagus increases the risk for esophageal cancer. The diagram below illustrates the continuum of cellular changes in the progression of Barrett's esophagus to invasive carcinoma. The actual risk is low. Estimates are that a gastroenterologist would have to do yearly endoscopic biopsies on 50 patients for 10 years to detect one cancer. And even then the cancer would be diagnosed early when it could be cured.

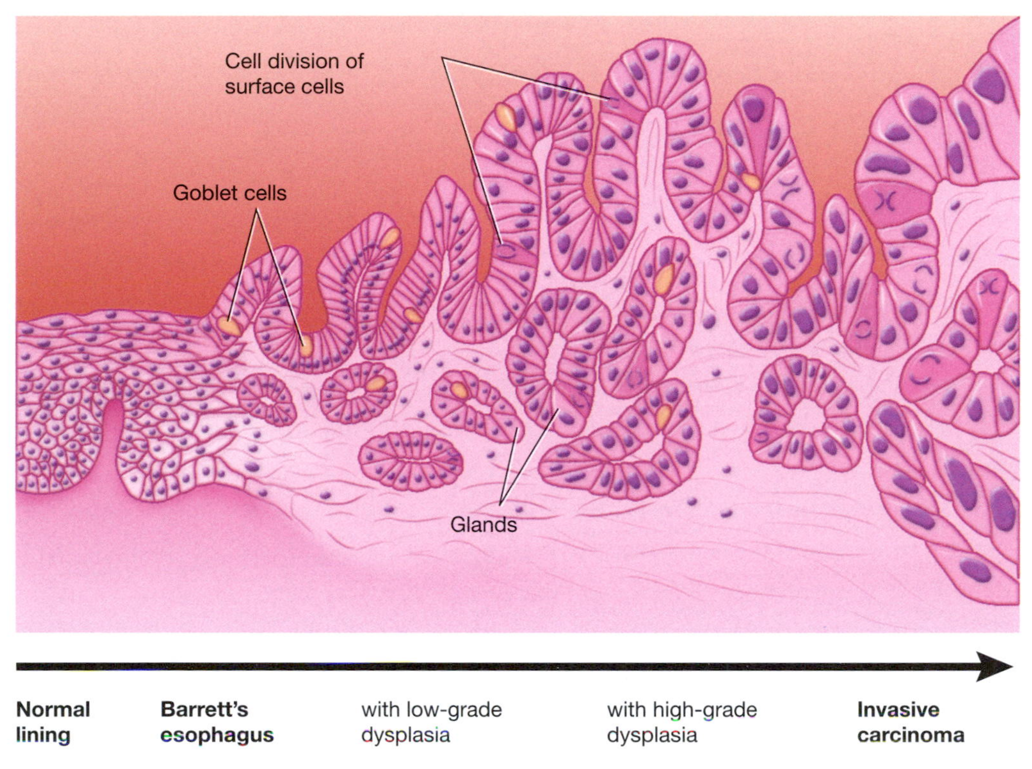

Courtesy of Johns Hopkins Pathology: http://pathology2.jhu.edu/bweb

6 *How will Mr. D. be treated?*

Fortunately, Mr. D. has early, low-grade dysplasia, so his risk of developing esophageal cancer is low. The only way to be sure he does not develop cancer cells, however, is to undergo periodic endoscopic biopsy surveillance. Surgery or other endoscopic therapies are reserved for patients who have high-grade dysplasia or actual cancer. Mr. D. will come back in 3 months for a second endoscopic biopsy and then once a year thereafter until no cells show dysplasia. Biopsies will be taken from four quadrant sites every 2 cm over the area of specialized intestinal metaplasia, as shown in the illustration shown here.

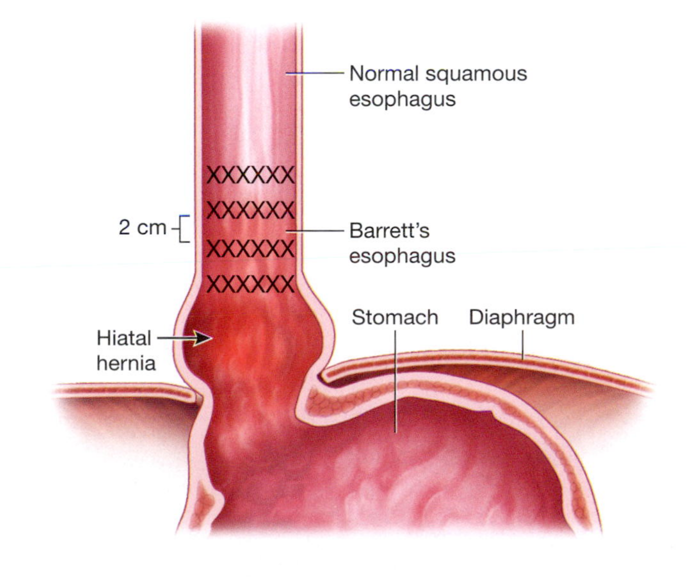

Most patients with Barrett's esophagus characterized by low-grade dysplasia do not go on to develop cancer. Also, no drug treatments have been proved to reduce the risk of esophageal cancer in patients diagnosed with low-grade dysplasia. Treatment for Mr. D. will be the same as for any other patient with GERD who does not have Barrett's esophagus:

a) lifestyle changes, and
b) medications to control heartburn symptoms.

Mr. D. will be given a prescription for a proton-pump inhibitor to reduce his gastric acid. Because calcium channel blockers used to treat hypertension and angina (Mr. D. takes amlodipine, Norvasc) have been associated with GERD symptoms, his antihypertensive medication is going to be changed. Mr. D. will be given an intramuscular vitamin B_{12} shot to replenish his depleted vitamin stores and will start taking a vitamin B_{12} supplement as recommended by the National Academy of Sciences for all Americans age 51 and older. Supplemental vitamin B_{12} is synthetic and not bound to protein, so it is highly bioavailable (well absorbed).

7 *What lifestyle changes can help to improve Mr. D.'s symptoms?*

Lifestyle changes that reduce the symptoms of GERD include the following.

- Stop smoking.
- Lose weight if overweight.
- Avoid high-fat meals and snacks, as a low-fat diet encourages gastric emptying.
- Elevate the head of the bed 6 inches with blocks under the bedframe, or use a full-length foam wedge to elevate the head. Use of pillows is insufficient.
- Consume a small last meal before bedtime and no food or drink at least 3 hours before bedtime.
- Take all medications with a full glass of water to avoid injury from medication directly contacting the esophagus.
- Consume a diet high in fruits, vegetables, and whole grains to reduce risk of cancer in general.
- Avoid foods that can cause heartburn by relaxing the esophageal sphincter, such as chocolate and peppermint.
- If a food seems to cause heartburn (such as corn dogs, in Mr. D.'s case), avoid them in the future.

After Mr. D. talks with his gastroenterologist about suggested lifestyle changes, he says, "My wife would kill me if she knew this, but I did start smoking again after quitting 20 years ago. I don't eat corn dogs that often, but I do eat a whole box of dark chocolate peppermint patties on every shift. It hides my cigarette mouth, and I thought dark chocolate was good for the heart!"

Mr. D. retired for good. He was able to quit smoking, and he became a volunteer for a middle-school conflict negotiation program. He has lost 20 pounds by volunteering with a hiking trails maintenance group. His last LDL level was 111 mg/dL and his HDL was 61 mg/dL. His RBC and MCV were normal. He continues to take a synthetic vitamin B_{12} supplement. Peppermint patties are a snack of the past. His last endoscopic biopsy showed that the areas of dysplasia were regressing and being replaced with normal esophageal epithelia tissue.

References

Barrett's Esophagus Information. Barrettsinfo.com
http://www.barrettsinfo.com/content/3a_what_is_egd_with_biopsy.htm

Buttar, N.S. & Wang, K. 2004. Mechanisms of disease: Carcinogenesis in Barrett's Esophagus. *Nature Clinical Practice Gastroenterology & Hepatology,* Dec. 27.
http://www.medscape.com/viewarticle/495345 (Membership to Medscape.com is free)

Institute of Medicine 1998. Dietary reference intakes for thiamin, riboflavin, niacin, vitamin B_{12}, pantothenic acid, biotin and choline. A Report of the Standing Committee on the Scientific Evaluation of Dietary Reference Intakes and its Panel on Folate, Other B Vitamins, and Choline and Subcommittee on Upper Reference Levels of Nutrients, Food and Nutrition Board, Institute of Medicine. Page 306. Washington, DC: National Academy of Sciences.
http://books.nap.edu/openbook.php?record_id=6015&page=306

18

Grandma's Got a Brand New Bag

Luanne Lin, age 83, was brought into the emergency department (ED) with upper gastrointestinal bleeding. She said she did not have any pain, but "all of a sudden I coughed and then I started to choke up bright red blood." By the time she reached the hospital, the bleeding had stopped. Her blood test results showed no abnormalities in coagulation. Her red blood cell (RBC) count was 3.5 K/cmm (normal: 4.00–5.20 K/cmm), but all other values were in the normal range. Her blood pressure was 110/70 supine and 100/60 standing. In Ms. Lin's purse was a pack of cigarettes, and in her wallet was a list of her prescription medications:

spironolactone/HCTZ, 25 mg/25 mg: Take in a.m. for hypertension
sertraline (generic Zoloft), 25 mg: Take in a.m. for depression
OTC nicotinic acid, 50 mg: Take in p.m. for high triglycerides

"The other pills I take are in a bag that my daughter is bringing in. I can't remember just now what else I take—I feel so lightheaded!"

Ms. Lin was given an IV infusion of normal saline. She was prepared for an endoscopic exam of her esophagus, stomach, and duodenum—an upper GI. A topical anesthetic was sprayed into the back of her throat, and she was given an IV sedative (propofol) to relax her and suppress her gag reflex. The gastroenterologist then threaded the endoscope (a thin, flexible plastic tube equipped with a miniature camera) into Ms. Lin's mouth, down her esophagus to her stomach and duodenum.

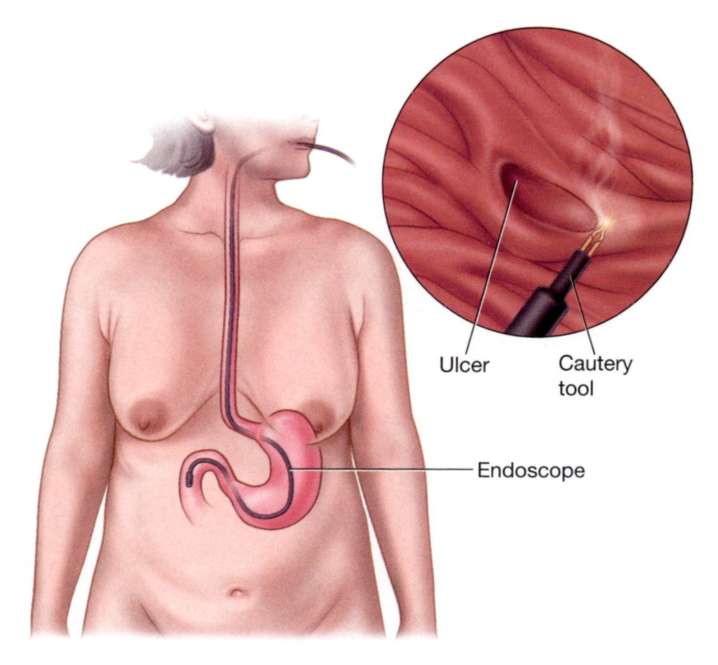

Below is a photograph of the only abnormality seen with the endoscope. Because there was no active bleeding or clot, the area was not treated. A biopsy sample was removed for further analysis.

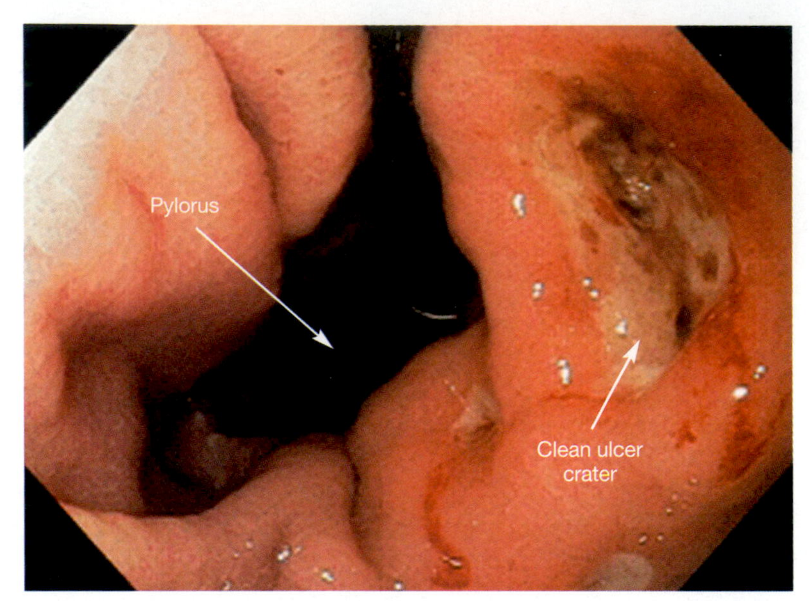

Courtesy of The American College of Gastroenterology

A photomicrograph of the gastric biopsy specimen under high power.

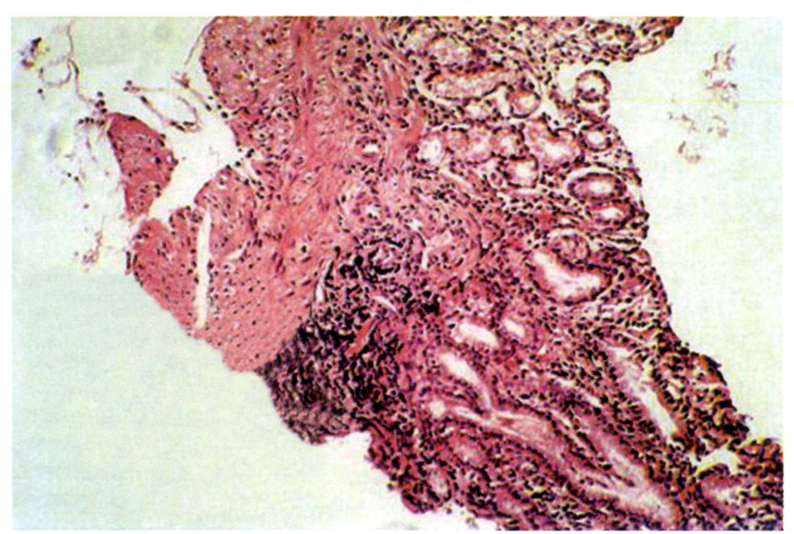

Questions

Name ______________________________ Section __________

1 Why was Ms. Lin given an IV infusion of normal saline?

2 What problem does Ms. Lin have?

3 What is the cause of Ms. Lin's problem?

4 What medications and self-medication behaviors can cause/aggravate her problem?

5 How will Ms. Lin's problem be treated?

Answers to Questions

1 *Why was Ms. Lin given an IV infusion of normal saline?*

Ms. Lin's RBC count is down, but her individual RBCs are normal (normal mean corpuscular volume, normal mean corpuscular hemoglobin concentration). This indicates anemia from acute blood loss. The normal saline will treat hypovolemia and restore her blood volume. This will help reduce her orthostatic hypotension (low blood pressure upon standing) and feelings of being "lightheaded." Normal saline is an isotonic fluid with a total osmolality equal to that of blood. It takes 3 L of an isotonic fluid to replace 1 L of blood because isotonic fluids also distribute to the extracellular fluid compartment.

Other isotonic fluids include lactated Ringer's solution and D_5W. Lactated Ringer's solution contains potassium and calcium, as well as sodium chloride. It is used to correct dehydration with hyponatremia (sodium depletion). D_5W contains glucose. It is not used to restore fluid volume because it can cause hyperglycemia. It is isotonic only until the glucose is metabolized, and then it becomes a hypotonic fluid. D_5W is used to correct increased serum osmolality.

IV influsions of normal saline can cause fluid volume overload (in the blood and extracellular spaces), which can result in hypertension and heart failure. Ms. Lin was monitored carefully for signs of fluid volume overload: tachycardia, increased blood pressure, distended neck veins, and abnormal lung sounds (crackles) with dyspnea (shortness of breath).

2 *What problem does Ms. Lin have?*

Ms. Lin has a gastric ulcer. The biopsy shows inflammation and gram-negative bacteria. There is no sign of cancer. An electron micrograph of the bacteria indicates that she is infected with *Helicobacter pylori.* This was confirmed with a fecal antigen test for the organism.

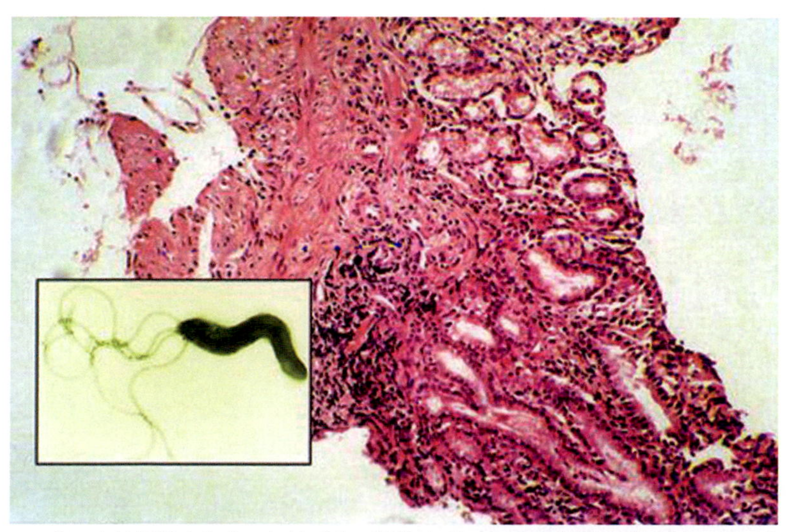

3 *What is the cause of Ms. Lin's problem?*

Helicobacter pylori plays a role in approximately 70% of gastric ulcers in the United States. *H. pylori* increases gastric acid secretion and infection/inflammation from the organism causes mucosal damage. The spiral organism penetrates the mucous layer and attaches to the epithelial layer. There, it releases phospholipase, which disrupts epithelial cell membranes and proteases, which break down epithelial cell proteins. Some strains of *H. pylori* have a cytotoxin-associated gene called *cag* A. This cytotoxin is associated with gastric cancer.

4 *What medications and self-medication behaviors can cause/aggravate her problem?*

Cigarette smoking increases the risk of infection with *H. pylori* and reduces the efficacy of antibiotic therapy to eradicate *H. pylori*. Non-steroidal anti-inflammatory drugs (NSAIDs), known as "aspirin-like drugs," inhibit synthesis of prostaglandins in the stomach that, in turn, are responsible for stimulating the synthesis of protective mucus. Examples of NSAIDs are aspirin, ibuprofen (as in Advil and Motrin), and naproxen (as in Aleve). NSAIDs at *any dose* and *any route of administration* carry increased risk of gastric ulcer. Other medications and self-medication behaviors that can increase the risk of gastric ulcer include:

- spironolactone (a potassium-sparing diuretic commonly used with thiazide diuretics to treat hypertension)
- nicotinic acid (a drug form of niacin that reduces triglycerides)
- alcohol (especially when taken within 2 hours after an NSAID)
- selective serotonin reuptake blockers (SSRIs) (a type of antidepressant)
- taking an NSAID and low-dose aspirin concurrently
- taking an NSAID and acetaminophen concurrently
- taking an SSRI and NSAID or acetaminophen concurrently

Ms. Lin's daughter arrived with her mother's "med bag." The daughter explained, "This is new. See how each pill bottle nicely fits in its own little compartment?" In the bag were the medications listed on the paper in Ms. Lin's wallet:

- spironolactone/HCTZ
- sertraline (generic for Zoloft, an SSRI)
- nicotinic acid

It also contained:

- Tylenol PM—extra strength (contains 500 mg acetaminophen and an antihistamine)
- ibuprofen (an OTC NSAID)
- low-dose aspirin
- omeprazole (an OTC proton-pump inhibitor (PPI) version of Prilosec used to reduce gastric acid)

Ms. Lin says she was starting to take the omeprazole 2 days ago to help with her "indigestion." She takes two ibuprofen every day when she wakes up. She also takes two Tylenol PM before her nap at 2 p.m. and two more before she goes to sleep at 9 p.m. She says she takes a low-dose aspirin before dinner ("for my heart"), along with a glass of red wine, sometimes two ("for my heart"). The combination of these medications likely played a role in the development of her gastric ulcer.

Following are examples of estimated relative risks (RR) of upper GI bleeding (compared to no drug):

- no drugs: RR = 1
- low-dose aspirin: RR = 1.5 – 3
- SSRI: RR = 2.6 – 3.6

- OTC NSAID: RR = 3.6
- acetaminophen (as in Tylenol), less than 2 grams/day: RR = 1.3
- acetaminophen (as in Tylenol), more than 2 grams/day: RR = 3.7
- OTC NSAID + low dose aspirin: RR = 5.6 – 7.7
- SSRI + low-dose aspirin: RR = 5.2
- SSRI + OTC NSAID: RR = 12
- omeprazole PPI + OTC NSAID: RR = 15
- OTC NSAID + 2 or more grams/day of acetaminophen: RR = 16

Ms. Lin is at greatly increased risk for a gastric ulcer because she is a smoker, has *H. pylori* infection, and takes prescription and OTC combinations of medications associated with upper GI bleeding.

5 *How will Ms. Lin's problem be treated?*

The goal for Ms. Lin will be to heal the ulcer, eradicate *H. pylori* infection, and prevent recurrence. Ms. Lin agrees to "stop smoking for good." She had been in a smoking cessation program, and her current use is two cigarettes per day, down from the pack-a-day habit she has had for 60 years.

Ms. Lin continues to take her SSRI and her antihypertensive medication. Her nicotinic acid is switched to a prescription version called Niaspan, which does not increase the risk of gastric ulcer. She will not take low-dose aspirin or pain relievers while she undergoes treatment.

She is given IV pantoprazole (Prontonix), a proton-pump inhibitor (PPI), to reduce gastric acid and help heal the ulcer. Omeprazole (as in Prilosec) is not prescribed because this PPI inhibits the metabolism of many other drugs, including the SSRI. Omeprazole also has been associated with increased risk of GI bleeding when taken with an NSAID. Ms. Lin is discharged on the oral PPI esomeprazole (Nexium), which does not interact with any of her other medications.

To eradicate *H. pylori* infection, Ms. Lin is given a "triple" antibiotic regimen of metronidazole, tetracycline, and a bismuth salt for 4 weeks. Ms. Lin must not drink any alcohol during her antibiotic treatment or the metronidazole will cause her to have a "disulfiram-like" reaction (also known as an "Antabuse reaction"), characterized by severe flushing and nausea. The bismuth will turn her tongue and stools black, and she is informed of this side effect. Ms. Lin also is given 200 mg lactoferrin and probiotic tablets twice a day for 4 weeks.

The combination of the proton-pump inhibitor/antibiotic treatment is known to produce an *H. Pylori* eradication rate of 80%. Addition of lactoferrin, a milk protein that binds iron, and probiotic supplements (containing lactobacilli and bifidobacteria) has been shown to increase eradication rates to nearly 89% while at the same time reducing antibiotic-associated GI side effects (nausea, diarrhea, abdominal pain).

After 8 weeks Ms. Lin feels fine. She provides a stool sample, which is tested for *H. pylori* stool antigen. The antigen test comes back negative, indicating that the *H. pylori* infection has been eradicated. A repeat endoscopy confirms the healing of her gastric ulcer and definitely rules out gastric cancer. She is advised to resume her low-dose aspirin, as evidence now suggests that a PPI with low-dose aspirin is more effective than switching from low-dose aspirin to clopidogrel (Plavix) in preventing recurring gastric ulcer bleeding. Ms. Lin is advised to avoid taking the OTC ibuprofen. OTC NSAIDs are more ulcerogenic when taken with low-dose aspirin and also prevent the beneficial effects of low-dose aspirin on platelet aggregation and they counteract the efficacy of antihypertensive medications.

PPIs decrease both calcium and vitamin B_{12} absorption from food. This is because gastric acid is needed to remove the nutrients from food proteins. Ms. Lin now will take a calcium supplement and a multivitamin with B_{12}.

Ms. Lin states that "Grandma's got a brand new bag. But I don't have the same stuff in it any more! I've finally stopped smoking. I've tossed the ibuprofen. I didn't really need it anyway. And it doesn't matter whether

I sleep during my nap, so I'll stop taking Tylenol PM then. I'll stick with Tylenol PM at night—it works for me. I'll take my low-dose aspirin at lunch and never with wine. If I get pain, I'll call my provider before I take anything. Acetaminophen seems to work for me, so maybe that's what she'll suggest."

References

Cerulli, M. (2006). Upper gastrointestinal bleeding. *eMedicine*, Feb. 3.
http://www.eMedicine.com/med/topic3565.htm

Santacroce, L. and Miragliotta, G. (2008). *Helicobacter Pylori* Infection. eMedicine, Aug. 14.
http://www.eMedicine.medscape.com/article/176938-overview

Shrestha, S. and Lau, D. (2007). Gastric ulcer. *eMedicine*, May 2.
http://www.eMedicine.com/MED/topic849.htm

Vega, C.P. (2007). Better treatment for *Helicobacter pylori*? A best evidence review.
Medscape Family Medicine, Nov. 5.
http://www.medscape.com/viewprogram/8097_pnt

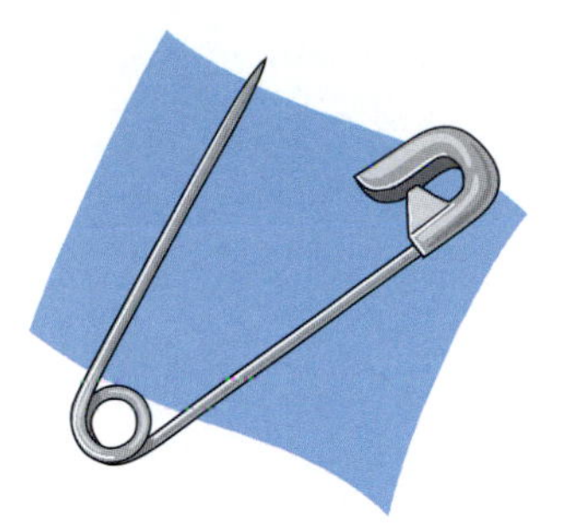

19

A Piercing Issue

Jessica Gangly, a 17-year-old, has a doctor's appointment at 8 a.m. She reports that her heart is "doing funny things." She has been attending dance camp and says she is fine in the morning but starts to feel "strange" by about 11 a.m. "My heart seems to skip beats, and then it stops for what seems like 5 seconds. Then it pounds. Sometimes I get these really rapid heartbeats and I see stars. My heart hurts, too—sort of a piercing pain. I'm doing dance camp because I want to model when I get out of high school. I'll never be a hand and foot model—I have man hands and flat feet—but I think I could do runway. Dance is helping me walk and move with some grace . . . um . . . which is sort of difficult at my height. I love clothes and design. Give me a few safety pins, and I can take just about any outfit and morph it into an edgy style."

Jessica is given a physical and an ECG. Selected results are given below:

Height: 6'0"
Weight: 120 lbs
BP: 110/67
Temp: 38° C
Lung ascultation: No abnormal findings
Heart ascultation: No abnormal findings. No mid-systolic click. No late systolic murmur.
ECG: Normal sinus rhythm.
ENT: Multiple conjunctival hemorrhages
Wears contact lenses.
Teeth, palate not remarkable; last dental exam 4 months ago
No lymphadenopathy
Family History: Not remarkable. Parents and both sets of grandparents are alive. No history of sudden cardiac death, stroke, or other cardiovascular disease, diabetes, or cancer. No history of thoracic aortic aneurism.
Patient is tall and thin with long fingers, long flat feet. Wears contact lenses and reports that she has a "–6.50" prescription. Patient complains of muscle aches, joint pain, and fatigue, which she attributes to long hours of dancing at camp.

Blood is drawn and sent to the lab. Jessica is scheduled at 11:30 for an echocardiogram, an ultrasound series of images of the heart obtained from the reflection or transmission of ultrasonic waves through cardiac (heart and aorta) tissue. After being asked if anyone in her family has ever been told that he or she has a thoracic aortic aneurysm, Jessica says she is beginning to feel very anxious. She says, "No one in my family—but a famous model dropped dead of it on the runway in Paris last year!"

Below is the report from her echocardiogram.

Echocardiography Laboratory

Name: Jessica Gangly

Age: 17

Findings

Indications: History of palpitations

ECG rhythm: Sinus rhythm with several PACs, PVCs

Study quality: Transthoracic M-Mode, Two Dimensional, Pulsed Doppler, Color Doppler, and Continuous Wave Doppler were performed. The study was technically adequate.

Left ventricle global: LV size, wall thickness, and systolic function are normal, with an EF > 60%

Right ventricle: The right ventricle is mildly enlarged, measuring between 31 and 35 mm. The right ventricular systolic function is normal.

Left atrium: The left atrium is normal in size.

Right atrium: The right atrium is normal in size.

Aortic valve: The aortic valve is trileaflet and appears structurally normal, with no aortic stenosis or regurgitation.

Mitral valve: The mitral valve is thickened with myxomatus degeneration. There is trace mitral regurgitation. There is mild thickening of the posteriol mitral valve leaflet. Mild prolapse of the anterior mitral valve leaflet.

Tricuspid valve: The tricuspid appears structurally normal. Trace/mild (physiologic) regurgitation.

Pulmonic valve: Pulmonic valve appears structurally normal, physiologic degree of pulmonic regurgitation.

Thrombus: There was no evidence of intracardiac thrombus.

Aorta: Limited views of the ascending aorta and the descending thoracic aorta are of normal caliber with no significant atherosclerotic disease or aneurismal dilatation.

IVC: The inferior vena cava is normal.

Pericardium: The pericardium is normal and there is no pericardial effusion.

Conclusions

The left ventricle is normal in size and function with EF greater than 60%.

There is no evidence of significant diastolic abnormality.

The aortic root size is normal.

There is no evidence of thoracic aortic dilation.

The right ventricle is mildly dilated.

There is no evidence of significant valve stenosis.

There is mild MVP with trace MR noted.

There is no pericardial effusion.

Questions

Name ______________________________ Section ________

1 What does a mid-systolic click and late systolic murmur indicate?

2 Why was an echocardiogram ordered if Jessica's physical exam was normal?

3 What is a thoracic aortic aneurism? What are the risks? How can it be treated?

4 What does Jessica's echocardiogram report show?

5 What can be done for Jessica's heart rhythm problem?

When Jessica's serology comes back from the lab, she is called to return immediately for an appointment for another blood culture. Both blood cultures, drawn 12 hours apart, are positive for *Staphylococcus aureus*. She undergoes another type of echocardiogram called transesophageal echocardiography (TEE). In this procedure, the patient is subjected to conscious sedation with a drug such as propofol (the patient is awake but will have no memory of the procedure). A local anesthetic is sprayed in the throat. The patient then lies on the left side and swallows the ultrasound probe until it is positioned in the esophagus directly behind the heart. A screenshot of the TEE follows.

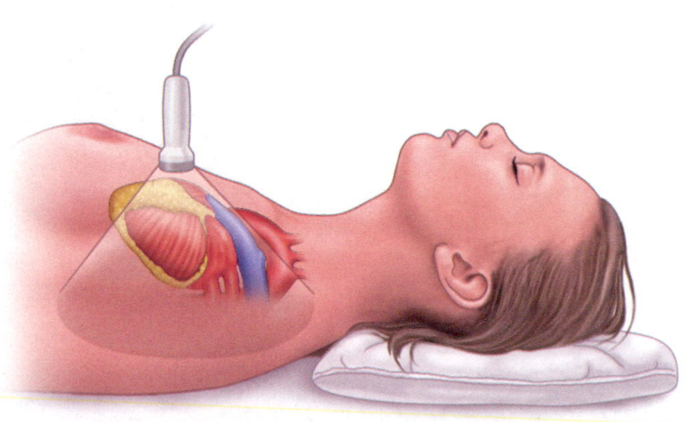

Standard or transthoracic echo

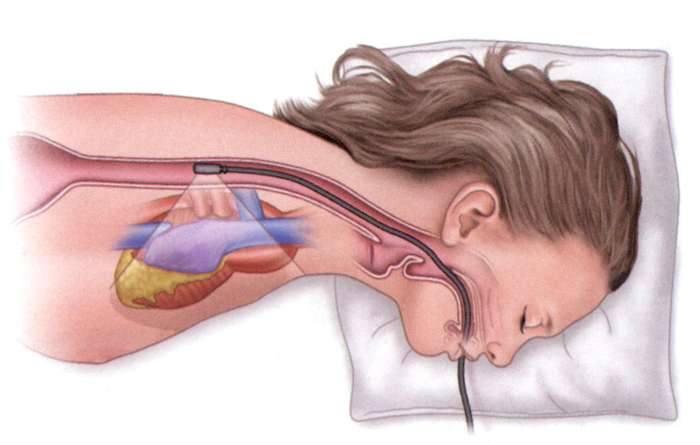

Transesophageal echo or TEE

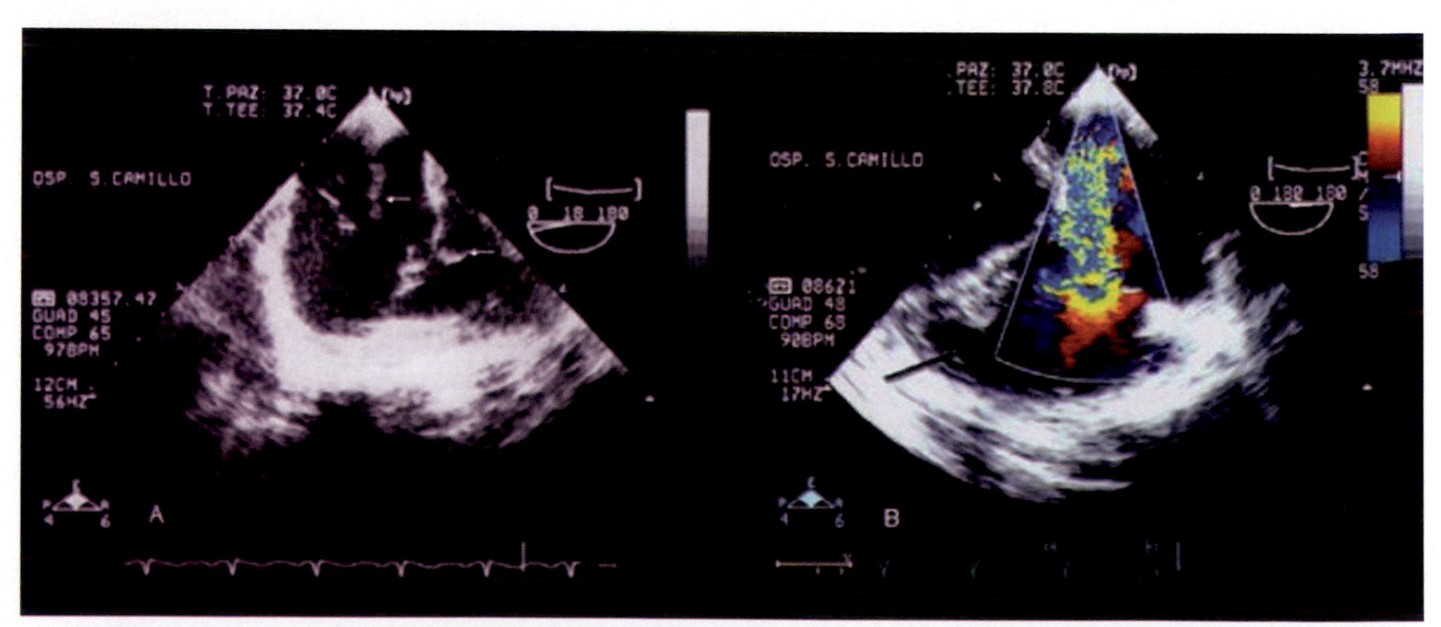

The arrows on the image on the left (A) point to two bacterial vegetations on the Eustachian and tricuspid valves.
The image on the right shows color Doppler imaging detecting regurgitation from the tricuspid valve toward the Eustachian valve.

6 What other problem does Jessica have, as shown on the transesophageal echocardiogram?

7 What was the likely cause of this problem?

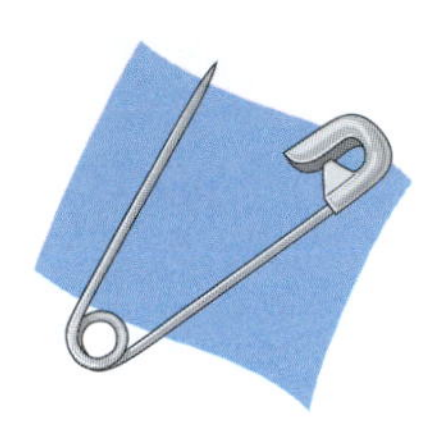

Answers to Questions

1 *What does a mid-systolic click and late systolic murmur indicate?*

Mitral valve prolapse (MVP) may be identified by auscultation when a mid-systolic click and late systolic murmur are detected. MVP is a common, often asymptomatic, finding in as much as 6% of the population. In the majority of MVP, a hemodynamic fluid volume, dysautonomia, is thought to be the underlying cause, rather than a defect in the mitral valve. MVP has been induced experimentally in healthy young adults by restricting fluid intake. Symptoms of MVP (also known as MVP syndrome) include palpitations, anxiety, panic attacks, migraine headache, and chest pain.

One or both of the mitral valve leaflets may be thickened and bulge into the left atrium during contraction. They may not close properly, and this allows blood to leak backward. The leaking blood (which is usually mild, or "trace") is known as regurgitation. Regurgitation causes turbulent blood flow and produces a murmur late in systole. The mid-systolic "click" is caused by the stretched mitral valve leaflets snapping against each other during contraction. MVP with regurgitation increases the risk of bacterial endocarditis, which can worsen mitral regurgitation. Small pieces of the bacterial "vegetations" can dislodge and cause a stroke.

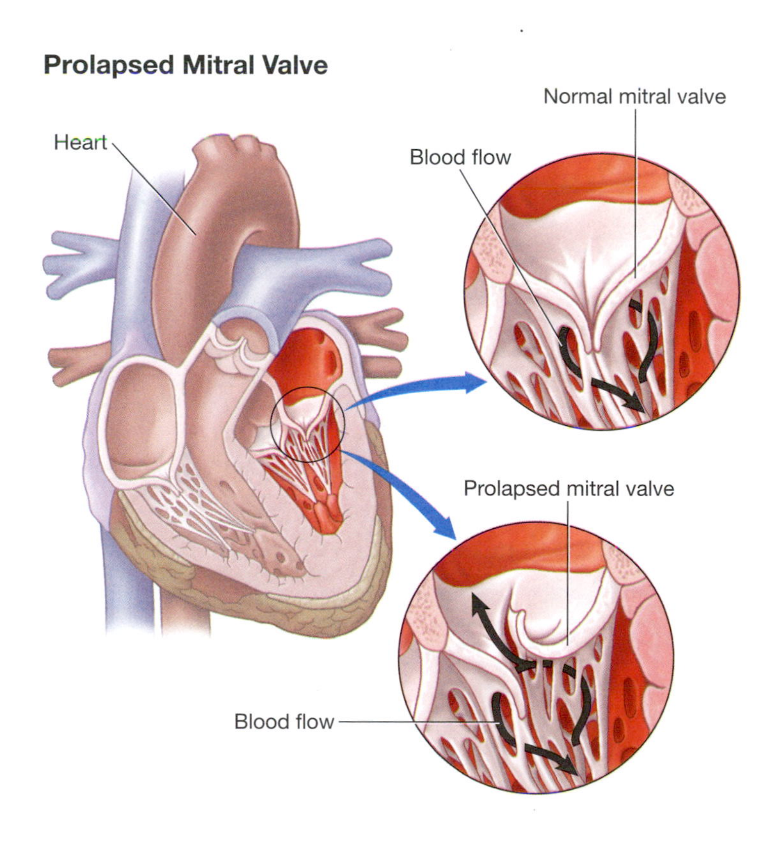

2 *Why was an echocardiogram ordered if Jessica's physical exam was normal?*

The American Heart Association (AHA) does not recommend an echocardiogram to screen for mitral valve prolapse when neither a mid-systolic click nor late systolic murmur is found with careful auscultation. Jessica, however, presents with five symptoms of Marfan's syndrome:

1. tall with thin bones,
2. long fingers,
3. flat feet,
4. severe myopia, and
5. self-report of palpitations (which could indicate MVP).

Other symptoms of Marfan's syndrome are:

1. a high arched palate and crowded teeth,
2. ectopia lentis (off center lens in the eye),
3. deep-set eyes,
4. long, thin face,
5. defects in the rib cage caused by overgrowth of the ribs that causes the sternum to either push out (pectus excavatum) or in (pectus carinatum),
6. joint laxity (loose joints) and scoliosis (curvature of the spine),
7. thoracic aortic aneurism, which can dissect (rupture).

Individuals with Marfan's syndrome have a mutation on chromosome 15, resulting in defective synthesis of fibrillin, a major protein in connective tissue. The disease has variable expression; thus, individuals have symptoms related to weakened connective tissue that vary from person to person.

Not all Marfan's syndrome is autosomal dominant. Although Jessica seems to have no family history, the disease can arise from a spontaneous mutation in about 25% of cases. Diagnosis of Marfan's syndrome is complicated if there is no family history and generally requires findings in at least three body symptoms. (Jessica has skeletal, cardiac, and vision symptoms; however, she may have inherited these from either parent.) Individuals with Marfan's syndrome often have MVP but, more important, they could have an aortic aneurism that can enlarge and suddenly rupture. An echocardiogram can detect a thoracic aortic aneurism.

3 *What is a thoracic aortic aneurism? What are the risks? How can it be treated?*

Thoracic aortic aneurism is an enlargement of the thoracic aorta, usually near the aortic root, the end closest to the aortic valve. The most common cause is atherosclerosis and uncontrolled hypertension, but it also is common in people with Marfan's syndrome. A ruptured thoracic aorta can cause sudden death in 90% of cases. When detected early and measured by echocardiogram and then verified by CT scan, patients can be monitored closely for change. Beta blockers, which reduce blood pressure and slow the heart rate, are usually prescribed. When the aneurism grows, and particularly when it reaches 5.0 cm or larger, surgical placement of a Dacron mesh graft in the aorta and replacement of the aortic valve with a mechanical valve can prevent acute aortic dissection. Because the artificial valve may stimulate the formation of blood clots, patients must take anticoagulants such as warfarin for the rest of their lives.

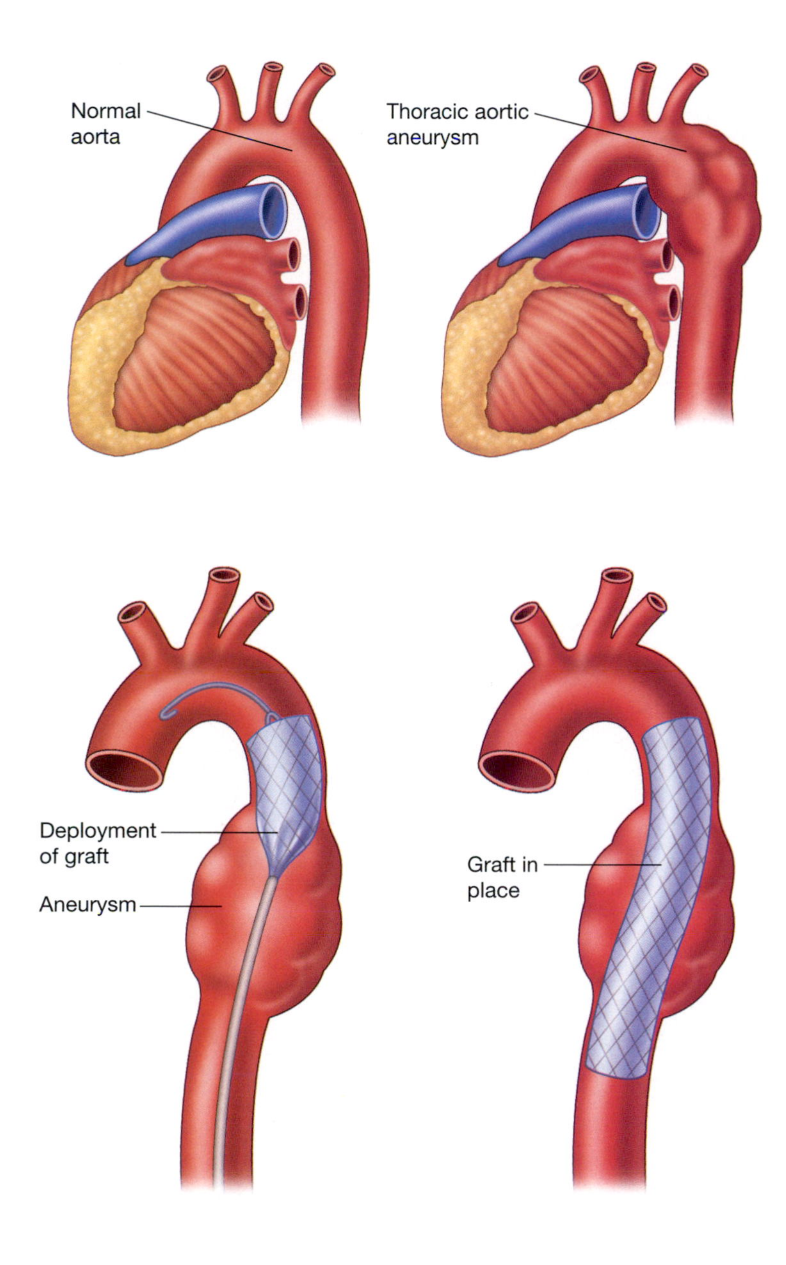

4 *What does Jessica's echocardiogram report show?*

An echocardiogram is a live action video of the heart and thoracic aorta by ultrasound. It allows the measurement of the aortic root, aortic arch, aorta, and left ventricular thickness, and it detects abnormal motion of the heart muscle, valve regurgitation (blood leaking backward through the valves), and stenosis (reduced blood flow through a valve). Jessica does not have a thoracic aortic aneurysm. She does have mitral valve prolapse with "trivial" regurgitation. She also has trace (mild) tricuspid valve regurgitation caused by mild dilation of the right ventricle—most likely secondary to mitral valve prolapse. Mild tricuspid valve regurgitation usually is considered benign and requires no treatment.

The associated ECG shows several premature atrial contractions (PACs) during the time Jessica noted her heart "skipping a beat." She also had several premature ventricular contractions (PVCs). PVCs and PACs are common in patients with MVP and typically are benign. The heart does not actually skip a beat; the extra beat comes sooner than normal, followed by a compensatory pause. These conditions are common in children and teenagers (even without MVP) and may occur more often in tall, thin, individuals who are volume-depleted. Jessica had not had anything to drink since 7 a.m., and by 11:30 she was volume-depleted—just as she was at dance camp during that time of day.

5 *What can be done for Jessica's heart rhythm problem?*

Maintaining adequate fluid and sodium intake and avoiding nicotine, caffeine, and oral decongestants (e.g., pseudoephedrine, phenylephrine), epinephrine in local dental anesthetics can help prevent rhythm disturbances and the associated anxiety in individuals with MVP syndrome. Often, cardiologists recommend an over-the-counter (OTC) once-daily magnesium supplement of 250 mg (66% of the magnesium RDA). Jessica also should avoid standing (which can trigger orthostatic hypotension and MVP symptoms). If the person must stand for periods of time, support stockings and contracting and relaxing the leg muscles periodically are helpful. Daily exercise (20 minutes to 2 hr/day) helps to downregulate cardiac beta receptors (exercise is sort of a natural beta blocker) so greater volume depletion or anxiety would be required to trigger dysrhythmia.

For most people with MVP syndrome, symptoms improve or even disappear with treatment. MVP does not shorten life expectancy. Previously, the AHA recommended that all patients with MVP with regurgitation to any extent take prophylatic antibiotics before dental procedures (including cleaning) or invasive procedures such as genitourinary or bowel procedures, bronchoscopy, or general surgery, to prevent infectious endocarditis (IE). The 2007 AHA guidelines, however, no longer recommend prophylactic antibiotics for healthy individuals with MVP because IE is extremely rare in this group and frequent exposure to anti-infectives (e.g., twice a year for dental cleaning) poses a much greater risk of producing drug-resistant bacterial flora.

6 *What other problem does Jessica have, as shown on the transesophageal echocardiogram?*

This time, TEE shows bacterial vegetations on the eustachian and tricuspid valves.

Jessica has a community-acquired *Staph. aureus* infection that has caused infective endocarditis (IE). TEE is highly sensitive and specific for IE. Her fever of 38° C, muscle aches and joint pain, and conjunctival hemorrhages are nonspecific symptoms of IE. IE involves the heart valve and/or the endocardium. Her tricuspid valve regurgitation may be caused in part by MVP and in part by IE on the valve.

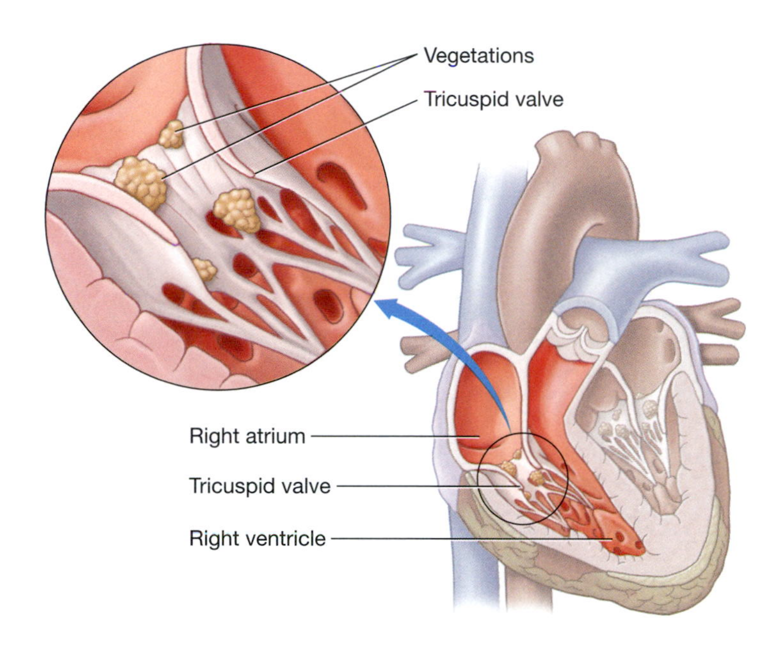

7 *What was the likely cause of this problem?*

Although dental cleaning was thought to increase the risk for IE in the past, recent research refutes this hypothesis (AHA statement) and the AHA no longer recommends prophylactic anti-infectives prior to dental procedures unless the individual is at high risk for IE (e.g., prosthetic heart valves, cardiac defects). Although MVP with regurgitation does increase the risk for bacterial endocarditis, the incidence is extremely rare. More than half of cases of IE are seen in males older than age 50 with prosthetic heart valves or cardiac defects such as atrial septal or ventricle defects. Elderly men are eight times more likely than elderly women to develop IE. In younger people, most cases of IE are in those with congenital heart defects, as well as those involving unsanitary IV drug use, tattoos, or piercings. Ironically, in Jessica's case the vegetations are not on the mitral valve and MVP may not have contributed to her IE. In fact, most cases of IE in young people involve the tricuspid valve.

When Jessica is questioned about IV drug use, tattoos, or piercings that might have provided a portal of entry for *Staph. aureus*, she says "Eww—no, not me!! Ummm . . . but I did have this safety pin that stuck into my back during a dance show. I had the girls alter the dowdy costumes by taking two tucks in the back of the tops. The girl who did my top stuck me in the back, and the pin went right through my skin. The show was about to start, so I just left it there. It didn't really hurt—not compared to my aching feet, that is!" An examination of the skin just behind her bra band revealed a small pustule.

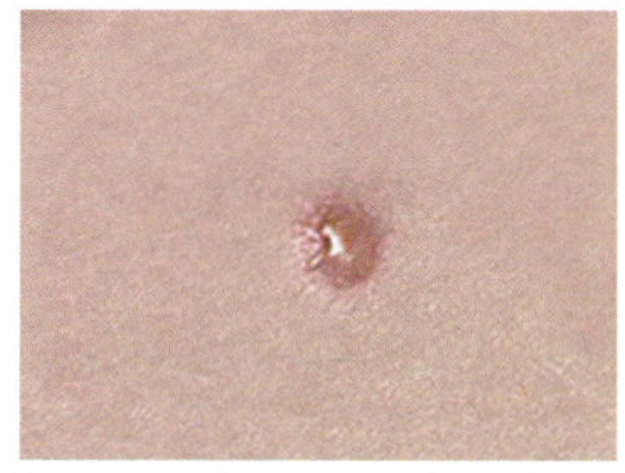

Image courtesy of www.dermnet.com

Fortunately, the *Staph. aureus* was not multiply drug resistant (MRSA), so the IE responded to anti-infective therapy. Though other people with MVP do not have to take prophylactic antibiotics, Jessica, because she has had IE, will have to take amoxicillin an hour before dental cleanings, invasive procedures, and surgery. Jessica now is a college student majoring in fashion design and merchandising. She still hopes to be involved with runway fashion—but on the "other side of the pin."

References

Marfan syndrome:
National Marfan Foundation:
http://www.marfan.org/nmf/index.jsp
http://www.io.com/~cortese/marfan/

Mitral valve prolapse:
Plewa, M.C., & Worthington, R. (2008). Mitral valve prolapse.
http://eMedicine.medscape.com/article/759004-overview

Dysautonomia:
http://www.mvprolapse.com/dysautonomia.html

American Heart Association prophylaxis guidelines:
http://www.americanheart.org/presenter.jhtml?identifier=3047083

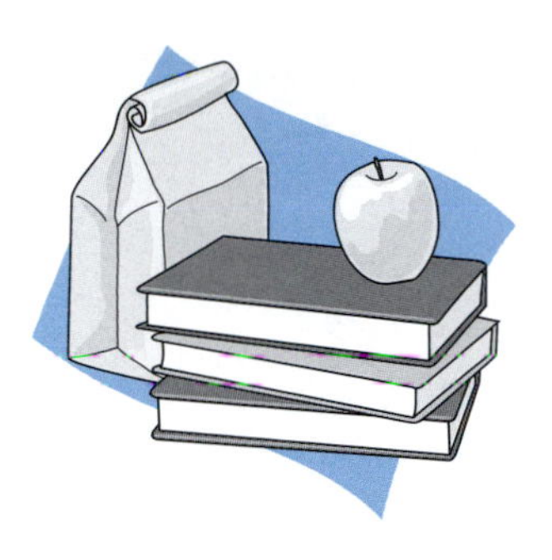

20

The Malicious Third Grade Cougher

During her well-child visit in mid-September, Mary was excited to be in the third grade. "We get to read chapter books and write stories. Plus—we have a pet hamster, and we take turns taking him home. My turn is next Friday!" Mary was weighed and measured and given a physical exam. All findings were normal for an 8-year-old girl. "I sure hope we don't have a repeat of last year," her mother sighed. "I lost all my sick leave staying home with poor Mary's colds. But she's been healthy all summer—and she grew two inches, too!"

On October 12 Mary caught her first cold. It came and went in 3 days over the weekend, leaving only a dry, hacking cough. Two weeks later Mary's teacher said to her mother at a parent–teacher meeting, "I think Mary is coughing in school on purpose. She doesn't seem to cough anything up. And from what you just said, she's okay at home. I'd like to make a referral to the school psychologist.

"Give me a week," Mary's mom responded. "I want to take her back to our nurse practitioner first."

Mary's APRN asked Mary to lean over, take a deep breath, and try to cough up sputum. Mary was unable to cough up anything. The APRN listened to Mary's heart and lungs and then told her mother that she would like Mary to have an evaluation at the Spirometry Lab.

Mary's spirometry readings below are before and after Mary was given 2 puffs of a short-acting bronchodilator called albuterol.

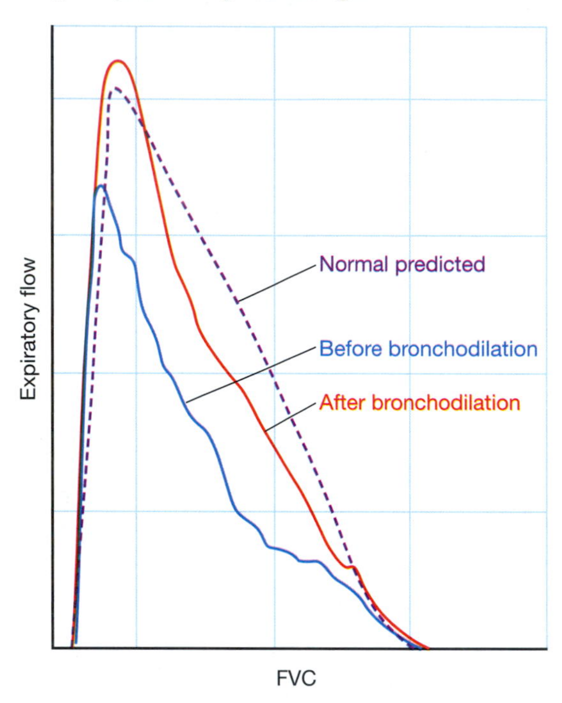

	Before Albuterol	After Albuterol	Normal (predicted)
FEV_1	1.59 L	2.04 L	2.09 L
FVC	2.57 L	2.64 L	2.44 L
FEV_1%FVC	62%	77%	86%

Sample data from: Become an Expert in Spirometry: http://spirxpert.com

Questions

Name ______________________ Section __________

1 What is spirometry?

2 What is FVC? What is FEV_1? What is $FEV_1\%FVC$?

3 What do Mary's spirometry readings suggest?

4 What problem does Mary most likely have? Why?

5 What is the value of a sputum culture in Mary's case?

6 What are the typical symptoms of this disease?

7 What are the probable triggers of Mary's cough?

8 How will Mary's cough be treated?

9 What is PEFR? How is it measured? What is it used for?

10 What other precautions should be taken for Mary?

Answers to Questions

1 *What is spirometry?*

Spirometry measures how fast a person can breathe in and out. It also measures the volume of air one can inhale and exhale.

Spirometry measures how fast and how much air you breathe out.

2 *What is FVC? What is FEV_? What is FEV_%FVC?*

FVC is forced vital capacity. This is the maximum volume of air (in liters) that an individual can rapidly force out of the lungs after taking a maximal inspiration.

FEV_1 is the forced expiratory velocity over the first 1 second of exhalation. This is the maximal volume of air that a person can forcefully expel in the first second after taking a maximal inspiration.

FEV_1%FVC is the FEV_1 to FVC ratio expressed as a percentage of the FVC. It has nothing to do with predicted values. The normal value is considered to be greater than 75%; that is, a person should be able to exhale at least three quarters of the total inspired in the first second.

3 *What do Mary's spirometry readings suggest?*

Mary's expiratory flow volume (as measured by the "area under the curve" of the graph) is less than normal. This suggests that she has bronchoconstriction. When she is given the bronchodilator, her expiratory flow improves, but not to normal levels.

Her forced vital capacity is normal (and actually greater than predicted). This indicates that her lungs are able to expel a normal volume of inspired air.

Mary's forced expiratory velocity over the first 1 second (FEV_1) is about 25% below normal. After the bronchodilator, her FEV_1 is close to normal.

Her $FEV_1\%FVC$ is only 62%, which is quite a bit lower than the minimal standard of 75% and the age-predicted standard of 86%. After the bronchodilator, Mary's $FEV_1\%FVC$ goes about 75% but is still below that predicted for her age.

4 *What problem does Mary most likely have? Why?*

Mary has bronchoconstriction, which can be treated with a bronchodilator. This suggests that she has asthma.

5 *What is the value of a sputum culture in Mary's case?*

Mary has a nonproductive cough without any fever or lung sounds indicative of infection. A sputum culture is unlikely to detect bacteria related to her cough.

6 *What are the typical symptoms of this disease?*

Typical symptoms of asthma in an 8-year-old child are wheezing and a sense of "difficulty breathing out." Many individuals with significant asthma-related bronchoconstriction, however, present only with a dry, nonproductive cough. The absence of a wheeze does not rule out asthma. In fact, a child who is wheezing and then suddenly stops wheezing may have developed complete bronchoconstriction and become critically ill in a matter of minutes.

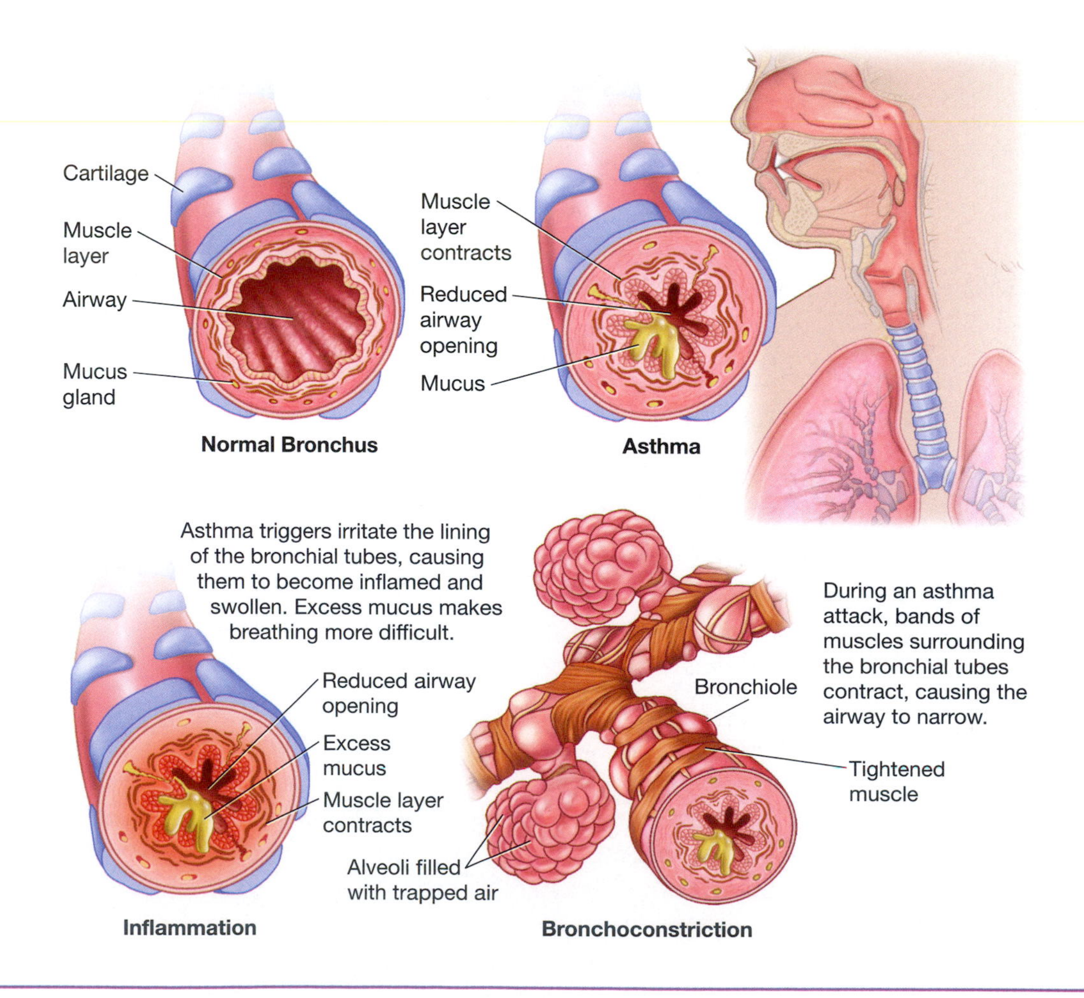

7 *What are the probable triggers of Mary's cough?*

Upper respiratory viruses (colds) are common triggers of cough/bronchoconstriction in children with asthma. And 30% of childhood asthma may be related to animal dander, particularly cats, but also to pet rodents such as mice, hamsters, and guinea pigs. Classrooms should not house such animals, and children with asthma should not be exposed to these animals.

8 *How will Mary's cough be treated?*

Mary will not be given a cough suppressant, because the underlying mechanism of asthma is inflammation of the airways. Many individuals with asthma also have airway hyper-responsiveness to various stimuli that act as triggers in the inflammatory process. Mary will be treated for her asthma with a daily inhaled corticosteroid to reduce her airway inflammation. Inhaled corticosteroids have minimal side-effects as long as they are used with a "spacer," a tube that attaches to the inhaler. This allows the patient to inhale the medication mist slowly and deeply into the lungs and helps to prevent droplets from settling in the throat. Mary will have to rinse out her mouth after each dose to wash away medication that remains in her mouth and throat. This will help to prevent the possibility of oral yeast infections, which can occur in patients taking inhaled corticosteroids.

Mary will be given a short-acting beta$_2$ agonist inhaler for acute episodes of tightness in the chest with or without cough or wheezing. When inhaled, the beta$_2$ agonist binds to beta$_2$ receptors in the lung, stimulating bronchodilation. Mary and her mother must understand that the doctor has to be contacted if Mary has symptoms more often than once or twice a week.

Inhaler With Spacer

9 *What is PEFR? How is it measured? What is it used for?*

Peak expiratory flow rate (PEFR) is a proxy (indicator) of FEV_1 measured using a handheld device. The patient takes a deep breath, then blows as rapidly as possible into the device. The device is calibrated to indicate "red," "yellow," and "green" zones like a traffic light.

Courtesy of Micro Medical Ltd., Quayside, Chatham, Maritime, Kent, ME4QY, United Kingdom

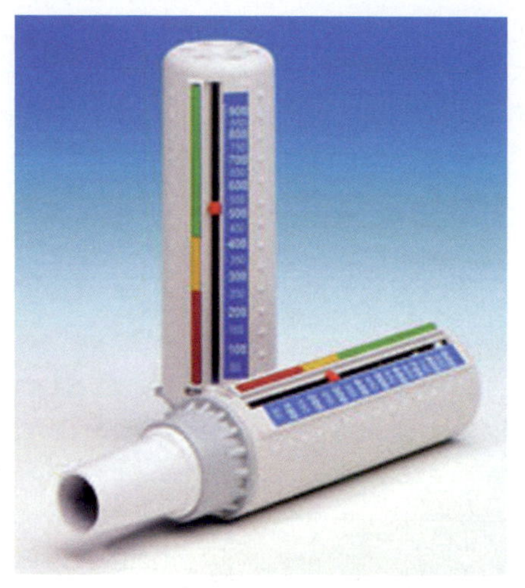

Courtesy of Micro Medical Ltd., Quayside, Chatham, Maritime, Kent, ME4QY, United Kingdom

GREEN (80% to 100% of personal best): If there are no symptoms, continue regular medicines and activities.

YELLOW (50% to 80% percent of personal best): Bronchoconstriction is occurring. Change or increase medications as directed by provider.

RED (below 50% of personal best): Significant bronchoconstriction. Use rescue inhaler as directed by provider. Contact provider! Recheck PEFR in 10–15 minutes. If PEFR improves, continue to monitor PEFR throughout the day. Provider may change daily or rescue medication type or dose.

Emergency Care: If PEFR does not improve, call 911! Do not attempt to drive patient to hospital!

10 *What other precautions will have to be taken for Mary?*

Mary also will be protected from common triggers such as animal dander and dust mites. The hamster was adopted by Mary's teacher, replaced by an ant farm and hermit crabs, which the children take turns bringing home. Mary's mother has encased her mattress and pillow in special protective linens to protect Mary from exposure to dust mites. Dust mites feed on skin particles that scale off humans and other mammals. Antigens in the excrement of dust mites trigger airway inflammation in susceptible people. If the relative humidity is kept below 40%, dust mites cannot survive.

A portable HEPA (high efficiency particulate air filter) has been placed in Mary's bedroom to clear the air of dust and mold spores. A portable dehumidifier has been put in her room to keep the humidity below 40%. Mary's mother also bought a vacuum cleaner equipped with a HEPA filter, and she vacuums after Mary goes to school. Because dust mites can accumulate on clothing and linens, Mary's mother washes her daughter's clothes in very hot water and dries them thoroughly. Mary is fortunate that her mother is able to undertake these modifications.

Mary contracts three more colds over the ensuing 4 months, but she does not cough and has not had to use the beta agonist as a rescue inhaler. Repeat spirometry tests suggest that the corticosteroid inhaler and environmental modifications are keeping Mary's asthma under good control.

References

Become an expert in spirometry
http://www.spirxpert.com

How to use a peak flow meter
http://patients.uptodate.com/topic.asp?file=al_asthm/2888

Case Studies in Environmental Medicine: Environmental Triggers of Asthma
http://www.atsdr.cdc.gov/HEC/CSEM/asthma/index.html

Guidelines for the diagnosis and management of asthma. (EPR-3)
National Heart, Lung and Blood Institute (NHLBI), July, 2007
http://www.nhlbi.nih.gov/guidelines/asthma/asthgdln.pdf

21

Favorite Uncle

Uncle Jed, the family's favorite uncle, choked on a turkey bone in his soup 3 days after Thanksgiving. He was able to cough up the little bone, but he began to hemorrhage from the esophagus. He was rushed to the emergency room, where he was intubated and a rapid influsion of 5% dextrose plus thiamine and colloid solution was started. He also was given an infusion of Sandostatin (octreotide acetate) to stop the bleeding. His blood was typed, and he received fresh frozen plasma, fresh blood, and vitamin K-1. He received endoscopic variceal ligation, in which elastic bands were placed around two ruptured varices (varicose veins) to strangle and obliterate them. He will require this procedure every 2–3 weeks until the esophageal varices have disappeared. He has been started on a beta blocker propranolol 40 mg PO bid and an ethanol drip. He has a nasogastric tube and is NPO for the next 24 hours. Gastric lavage through the nasogastric tube will be performed repeatedly over the next 24 hours to examine aspirated stomach contents and identify any rebleeding.

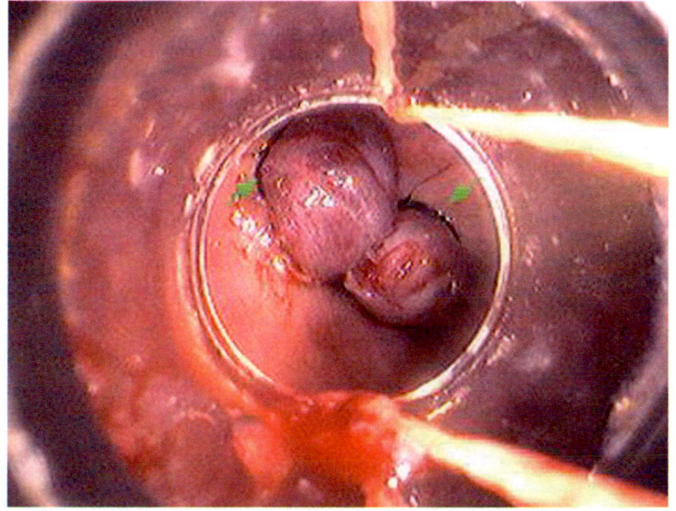

Reprinted with permission from: Runyon. Patient Information: Screening for esophageal varices. In: UpToDate, Basow, DS (Ed), UpToDate, Waltham, MA 2009. Copyright © 2009 UpToDate, Inc. For more information visit www.uptodate.com

Uncle Jed's niece Janice and nephew Greg just visited him. "Whoa!" said Greg. "Uncle Jed has man-boobs! But otherwise he looks okay. He does have a great tan."

"I don't know about that tan," Janice responded. "It looks pretty orange—probably fake! But he does have breasts—and stretch marks on his beer belly. Weird. Did you notice that he also has no hair on his arms or legs? He seems to have changed since last Thanksgiving—not just the tan thing either. He has red splotches on his nose and cheeks, and his palms are really red. But the thing I worry about the most is that he seems to lie about things. And not in a kidding way like when we were kids. I asked him which kennel his dog was in so I could call and tell them to keep Snarl another week. And Uncle Jed told me that Snarl was at home with Aunt Sally. Well, Aunt Sally died of breast cancer two years ago. Does Uncle Jed have Alzheimer's Disease?"

Following are selected lab values from Uncle Jed, 24 hours after admission.

Test	Value	Reference Range	Units
Ammonia	97	1960	mcg/dL
BUN	42	7–18	mg/dL
Creatinine, serum	1.52	0.8–1.3	mg/dL
Bilirubin, total	1.7	0.1–1.0	mg/dL
Bilirubin, direct (conjugated)	0.3	0.1–0.3	mg/dL
Bilirubin, indirect (unconjugated)	1.4	0.2–0.8	mg/dL
Protein, total	5.5	6.4–8.2	gm/dL
Albumin	2.4	3.4–5.0	gm/dL
Alkaline phosphatase	157	50–136	U/L
AST/SGOT	49	15–37	U/L
ALT/SGOT	71	30–65	U/L
GGT	139	1–94	U/L
Cholesterol, total	230	50–199	mg/dL
Triglycerides	301	15–149	mg/dL
RBC	4.1	4.50–5.90	K/cmm
HGB	13.5	13.5–17.5	g/dL
HCT	37	41.0–53.0	%
MCV	102	80–100	fL
MCH	26	26.0–34.0	pg
MCHC	31.0	31.0–37.0	g/dL
PT (prothrombin time)	14	9–12	seconds

Questions

Name ______________________________ Section __________

1 What health problem does Uncle Jed probably have?

2 What caused the esophageal varices?

3 Why would Uncle Jed be started on a beta blocker (which slows the heart and lowers blood pressure)?

4 Why would Uncle Jed be started on an ethanol drip?

5 Why was Uncle Jed given thiamine along with the dextrose infusion?

6 Why does Uncle Jed have breasts, stretch marks, a tan, and red marks on his face?

7 Uncle Jed is positive for hepatitis C. How does this viral infection play a part in his current problem and prognosis?

8 What other problems might Uncle Jed have?

9 Why is Uncle Jed telling lies?

10 What is Uncle Jed's prognosis? Are there any treatments for this?

Answers to Questions

1 *What health problem does Uncle Jed probably have?*

He has cirrhosis of the liver. All of his liver enzyme levels (ALT, AST, GGT, alkaline phosphase) are elevated. Below is a slide made from tissue taken during a liver biopsy. Notice how cirrhosis of the liver produces fibrous bands of tissue surrounding the liver nodules. Bile is prominent within the nodules.

Normal Liver Biopsy

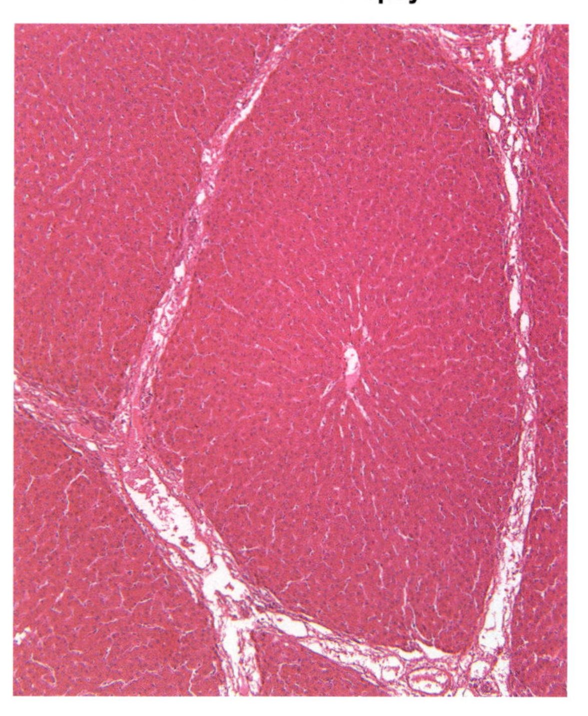

Cirrhosis Liver Biopsy

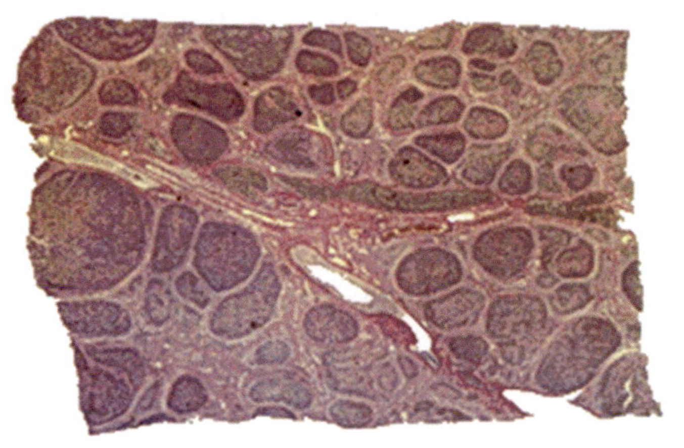

From Pathweb.UCHC.edu The Virtual Pathology Museum

2 *What caused the esophageal varices?*

Esophageal varices are actually varicose (swollen) veins. In cirrhosis of the liver, portal blood flow to the liver is impeded because of liver fibrosis. This causes portal hypertension and a shunting/backflow of blood into the veins in the esophagus. The liver does not synthesize clotting proteins well in individuals with cirrhosis. This adds to the risk of hemorrhage when esophageal varices rupture. To help his blood to clot, Uncle Jed was given fresh frozen plasma containing clotting proteins, vitamin K1, and Sandostatin (octreotide acetate).

Portal Hypertension

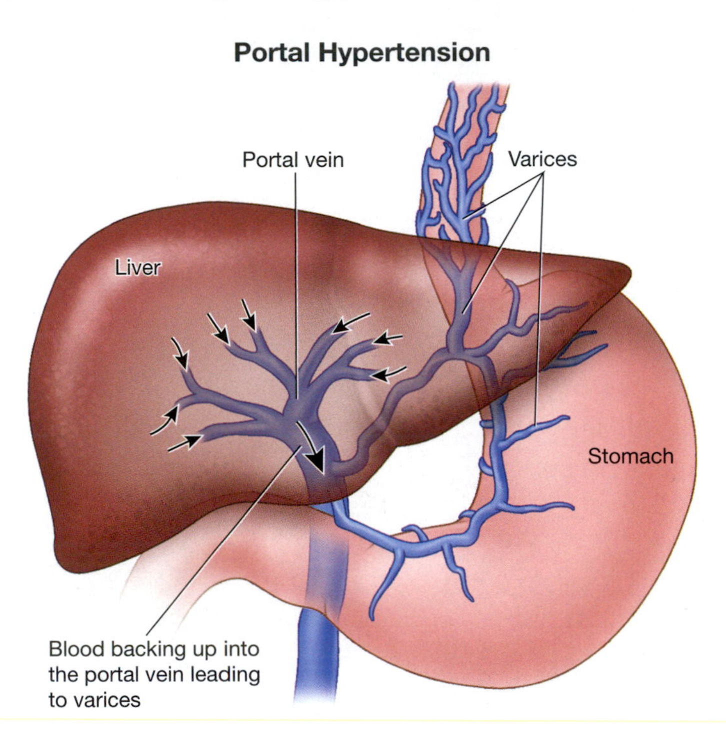

3 *Why would Uncle Jed be started on a beta blocker (which slows the heart and lowers blood pressure)?*

Reducing the resting heart rate by 25% (or decreasing the heart rate to 55 beats per minute) has been shown to reduce portal hypertension and reduce the risk of future ruptures of esophageal varices.

4 *Why would Uncle Jed be started on an ethanol drip?*

Uncle Jed revealed that he has "about four drinks a day—a scotch before dinner and usually two or three glasses of wine with dinner." The purpose of the ethanol drip is to prevent acute alcohol withdrawal, which could result in vomiting and a seizure—both of which could cause rebleeding. During his hospital stay the ethanol drip will be tapered slowly so Uncle Jed can withdraw from alcohol safely. Many hospitals use a benzodiazepine sedative such as lorazepam (Ativan) instead of ethanol for this purpose.

Four drinks a day puts Uncle Jed in the "moderate" alcohol intake category. But when the clinical dietitian came in with different sizes of wine glasses, the glass size that Uncle Jed said he uses actually holds 8 oz. A "standard drink" (see References, page 192) is 4 oz of wine containing 12% alcohol. Uncle Jed actually has been consuming seven standard drinks a day for many years—as did Aunt Sally. This is considered heavy drinking and is a risk for alcohol-induced cirrhosis. (Note: Heavy alcohol consumption in women is associated with an increased risk of breast cancer.)

5 *Why was Uncle Jed given thiamine along with the dextrose infusion?*

A rapid infusion of dextrose will exhaust thiamine reserves, which are low in heavy drinkers. The resulting acute onset of thiamine deficiency causes Wernicke encephalopathy with nystagmus, ataxia, and confusion. All patients with suspected heavy alcohol use should be given thiamine when dextrose is administered.

6 *Why does Uncle Jed have breasts, stretch marks, a tan, and red marks on his face?*

Alcohol-induced cirrhosis causes degeneration of liver tissue and, consequently, a reduction in all hepatic function. The breast tissue (gynecomastia) results from the inability of the liver to metabolize estrogen. Reduced hepatic albumin synthesis leads to reduced plasma albumin. An important function of plasma albumin is to keep water in the blood, called plasma oncotic pressure. Ascites—movement of water to the peritoneal cavity—occurs when there is insufficient plasma albumin, as in protein malnutrition and/or cirrhosis. The increase in portal pressure also contributes to ascites. Stretch marks, striae, occur as the collagen in the skin pulls apart to stretch over the fluid filled abdominal cavity.

The ability of the liver to conjugate bilirubin is reduced in cirrhosis, and jaundice results. The unconjugated bilirubin imparts an orange color to the skin, which in some skin types may look like a tan. Orange-colored sclera can confirm that the patient has jaundice and not a bad fake tan job. The red marks on Uncle Jed's face are spider angioma, arterioles that grow on the surface of the skin. These are common in heavy drinkers but also are seen in many older adults and in patients with rosacea, a disease now thought to be caused by skin mites and treatable with topical anti-infectives.

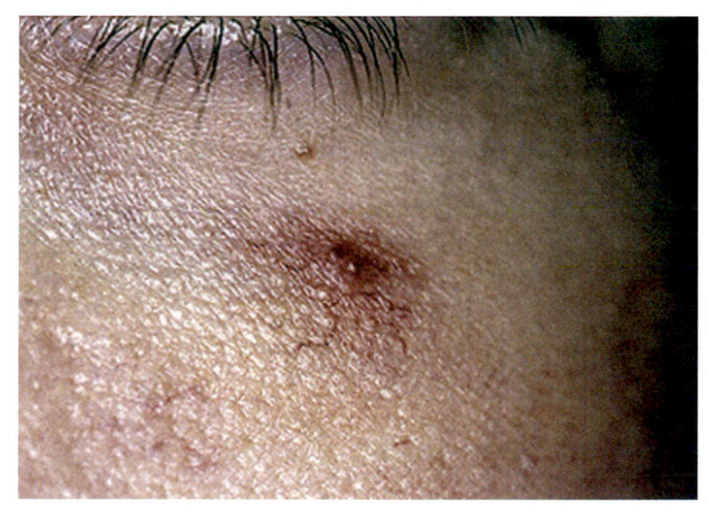

Permission granted from: TB Fitzpatrick, *Color Atlas and Synopsis of Clinical Dermatology*, 3 Ed, Copyright 1997, McGraw-Hill.

7 *Uncle Jed is positive for hepatitis C. How does this viral infection play a part in Uncle Jed's current problem and prognosis?*

Cirrhosis eventually develops in 20% or more of individuals who have had infection with hepatitis C for 20 years or more. Male gender and heavy alcohol consumption greatly increase the risk of a person with hepatitis C developing liver disease. The source of Uncle Jed's infection with hepatitis C is thought to be either a tattoo from contaminated equipment or a blood transfusion in the 1980s before the blood supply was screened for the virus starting in 1992. If Uncle Jed had not had hepatitis C, he might not have developed cirrhosis from his drinking pattern, as only about 10% of heavy drinkers develop cirrhosis.

8 *What other problems might Uncle Jed have?*

Other functions of the liver include synthesis of immunoglobulins, activation of vitamin D, and toxin and drug metabolism. Uncle Jed is at risk for infection, osteoporosis, and drug toxicity, all of which are more common in people with cirrhosis. Some of the portal blood that backs up in the esophageal veins is shunted to the systemic circulation—called a porta-caval shunt. Thus, material such as ammonia that the liver normally clears will concentrate in the systemic blood and cross the blood-brain barrier. The result is encephalopathy. Early symptoms of encephalopathy, such as nystagmus and flapping tremor in the hands, progress to stupor, coma, and death.

Uncle Jed also has hypertriglyceridemia caused by his heavy drinking. Alcohol is converted to triglycerides (fat) in the liver. This contributes to fatty liver and hypertriglyceridemia. Although one or two drinks a day can be good for the heart by increasing HDL, the "good" cholesterol, more than that causes high triglycerides, an independent risk factor for heart disease in men.

He also has anemia with a low hematocrit (HCT), reduced red blood cell count (RBC), and increased mean corpuscular volume (MCV). His hemoglobin concentration (HGB), mean corpuscular hemoglobin (MCH), and mean corpuscular hemoglobin concentration (MCHC) are normal. The anemia is secondary to bleeding and bone marrow suppression from chronic heavy alcohol intake. Also, alcohol impairs folate absorption and increases folate elimination. Folate deficiency slows red cell maturation; as a result, the mean corpuscular volume is larger than normal.

9 *Why is Uncle Jed telling lies?*

Cirrhosis can cause Korsakoff's psychosis, which affects recent memory and ability to express thoughts. The brain compensates by "filling in" the missing thoughts, known as confabulation. Studies suggest that an episode of Wernike's (acute thiamine deficiency) may trigger Korsakoff's psychosis. Hence the term Wernike-Korsakoff's syndrome. Intravenous thiamine is used to treat Korsakoff's psychosis, but it usually is ineffective in improving memory.

10 *What is Uncle Jed's prognosis? Are there any treatments for this?*

Unfortunately, the prognosis for a heavy drinker with hepatitis C related cirrhosis is poor unless the person is able to undergo a liver transplant. There are six genotypes of hepatitis C. The most difficult to treat is genotype 1, the most common type in the United States. Therapy with daily ribavirin (an antiviral drug) and weekly injections of pegylated interferon alpha have been shown recently to eliminate hepatitis C in 40% of patients with genotype 1 for up to 6 months following therapy. But these studies have been conducted with individuals before they develop cirrhosis. Side-effects of interferon therapy include flu-like symptoms, fatigue, nausea, vomiting, headaches, and reduced production of platelets (thrombocytopenia) and white blood cells. Side-effects of ribavirin include reduced production of red blood cells. Because of Uncle Jed's advanced cirrhosis, he is not offered interferon/ribavirin therapy.

Uncle Jed is scheduled to undergo a distal splenorenal shunt procedure in 6 weeks. Blood from the varices will drain into the splenic and gastric veins to the shunt. As a result, pressure will decrease in the portal vein and help to decrease the risk of esophageal bleeding in the future.

Uncle Jed does well after his surgery. His ascites improves. He continues to take propranolol and thiamine. He takes a diruretic (furosemide) and follows a low-sodium diet to help reduce his ascites. He no longer drinks any alcohol at all. As a result, his triglyceride level decreases. He follows a vegetarian diet that provides enough protein to meet his recommended dietary allowance (RDA). (Animal protein increases blood ammonia, the principal toxin in encephalopathy.)

He takes lactulose (a sugar that is not absorbed) to make the pH of his stool more acid. This helps to reduce the absorption of ammonia from the colon. He also takes *Lactobacillus acidophilus* capsules to help

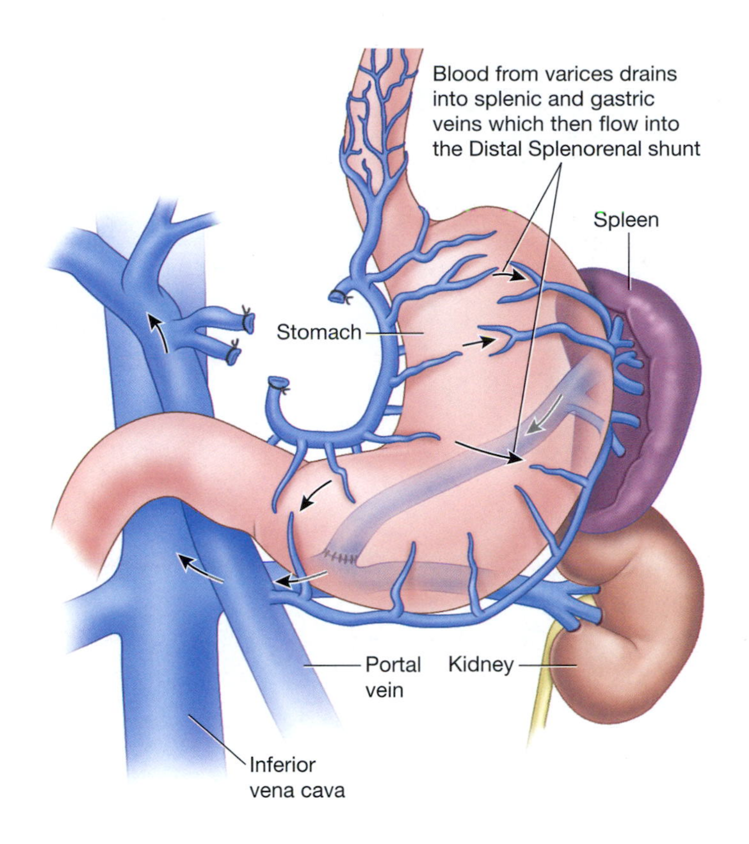

colonize his colon with non-ammonia-producing bacteria. He takes calcitriol (an active form of vitamin D) to help prevent bone loss. He takes a supplement of fish oil (omega-3 fatty acids have been shown to help protect the liver in cirrhosis) and a zinc supplement (shown in some studies to lower blood ammonia levels and help protect against encephalopathy), and he avoids all over-the-counter (OTC) medicines. He got a flu shot this year.

Against all odds, Uncle Jed attends the family Thanksgiving the following year. Tofu turkey is on the menu.

Normal Liver

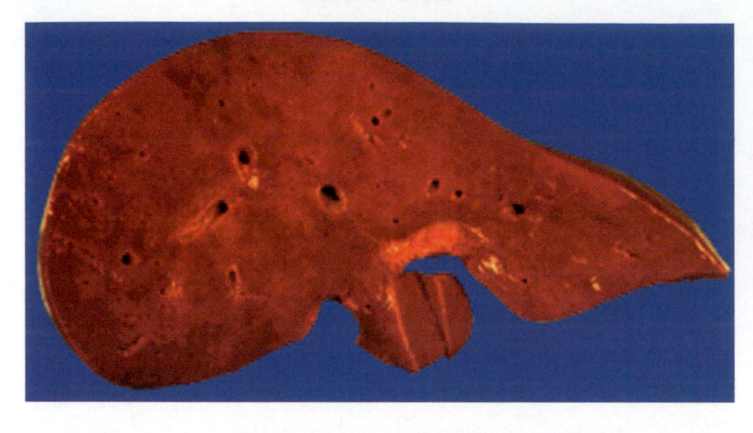

Normal Liver

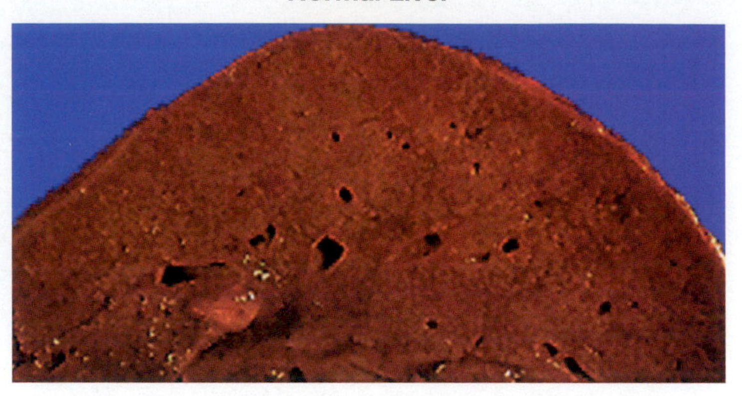

Liver With Cirrhosis

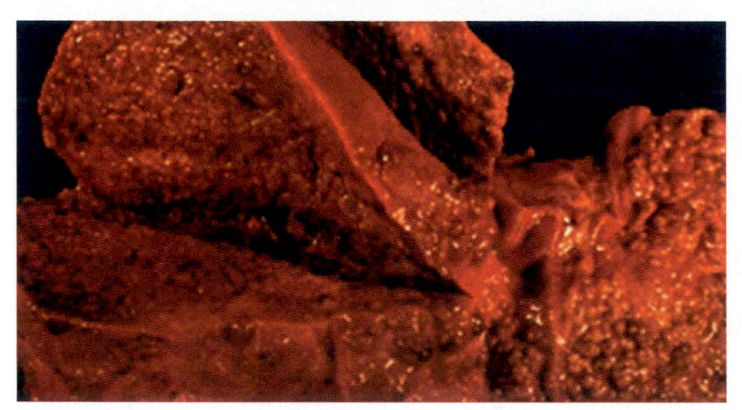

Liver With Cirrhosis

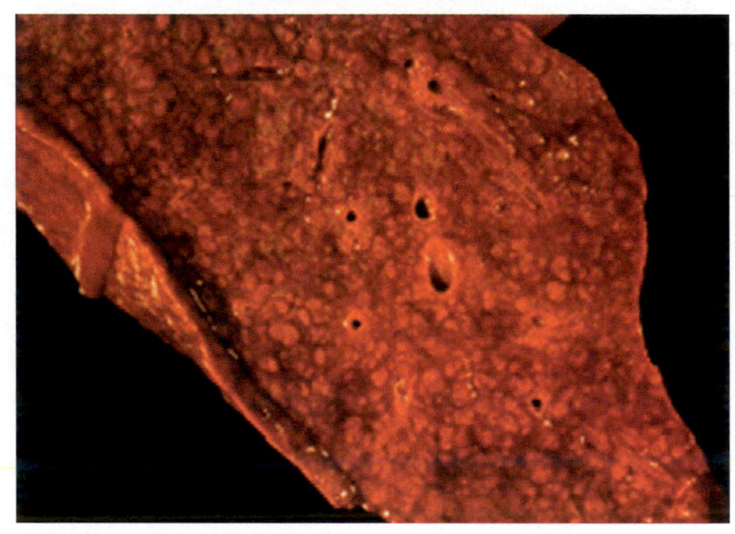

Photos from Pathweb.uchc.edu The Virtual Pathology Museum.

Note the light-colored nodules surrounded by fibrous bands (fibrosis).

References

Azer, A. (April 12, 2006). Esophageal varices, eMedicine. Accessed June 6, 2007, from:
http://www.eMedicine.com/med/topic745.htm

Marsano, L., Mendez, C., Hill, D., Barve, S., & McClain, C. (2003).
Diagnosis and treatment of alcoholic liver disease and its complications. Alcohol Research and Health 27(3), 247–256. Accessed March 25, 2009, from:
http://pubs.niaaa.nih.gov/publications/arh27-3/247-256.pdf

Standard drink conversion:
http://www.virginia.edu/case/ATOD/standard-drink-conversion.html

22

Never Too Old to Learn a Bad Habit

Olivia G., age 98, decided on her own to enter the Evencare Assisted Living complex. Other than hypothyroidism (kept in control with thyroid replacement hormone for the past 50 years) her only health problem is open angle glaucoma and macular degeneration. "Between the two, I really can't see much at all—especially the buttons on the phone and the microwave," she says, "but I'm lucky because I still have perfect memory, a good appetite (especially for dark chocolate!), and a sense of humor. Olivia settled in easily and formed a fast friendship with three sisters—"the Goodwin Girls"—20 years her junior.

A month after arriving, Olivia developed a fever of 99° F and an itching and burning sensation when she urinated. Her urine developed an orange color and had a strong odor.

Questions

Name ______________________________ Section ________

1 What is open angle glaucoma, and what effect does it have on vision?

2 What is macular degeneration, and what effect does it have on vision?

3 What is the most likely cause of Olivia's new problem—fever with itching and burning during urination?

4 How can this new problem be treated?

2 *What is macular degeneration, and what effect does it have on vision?'*

The macula, in the back of the eye, is essential for clear central vision. There are two types of macular degeneration—wet and dry. In dry macular degeneration the retina thins and develops deposits of debris called drusen. The macular regions on and near the drusen deposits degenerate, resulting in loss of central vision. Dry macular degeneration—also called age-related macular degeneration—accounts for 90% of the cases of macular degeneration.

In the less common wet macular degeneration, the drusen deposits cause new blood vessels to grow (neovascularization) under the retinal layer. These new blood vessels are fragile; they hemorrhage and quickly kill retinal cells.

Olivia has dry macular degeneration. Because she has lost a significant amount of central vision, she relies on her peripheral vision to negotiate her surroundings. She no longer can read but she does enjoy audio books.

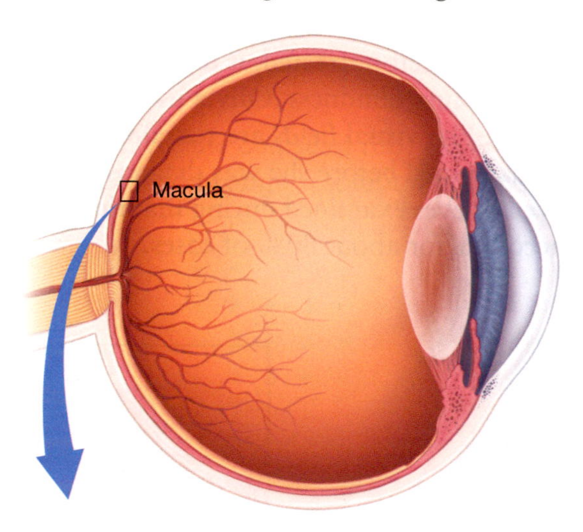

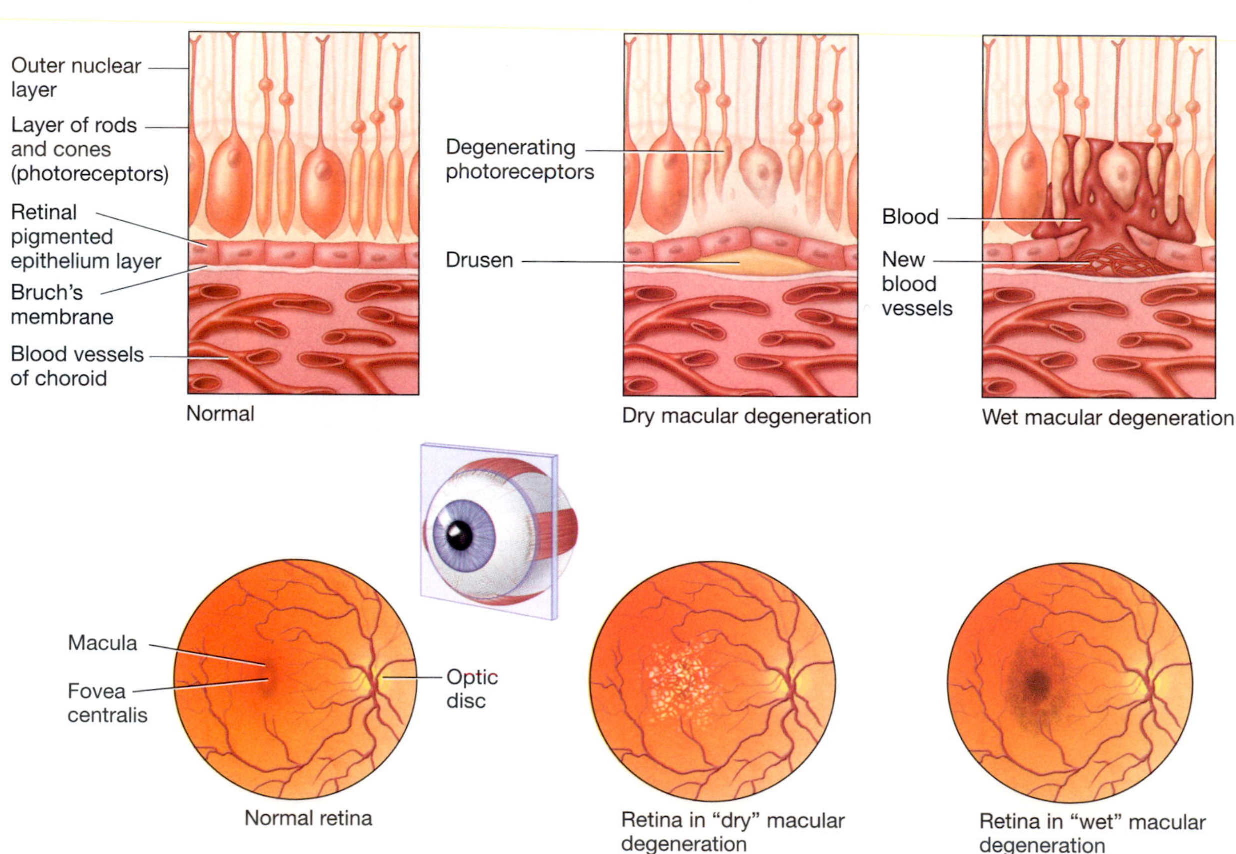

3 *What is the most likely cause of Olivia's new problem—fever with itching and burning during urination?*

Olivia has symptoms of a urinary tract infection (UTI). The most common cause is *Escherichia coli* (*E. coli*) bacteria, found in feces. Fecal bacteria can migrate from the rectum to the urethra. "Wiping from front to back" is recommended, but the distance from the rectum to the urethra and the length of the urethra is shorter in women than men, so fecal bacteria have an easier time gaining entry to the bladder in women. This is why women have UTIs more often than men do. Sexually active women have more UTIs because intercourse, spermicidal agents, and diaphragms can irritate the urethra. Estrogen loss in post-menopausal women causes the epithelial tissues of the vagina, urethra, and bladder to thin and become more susceptible to infection.

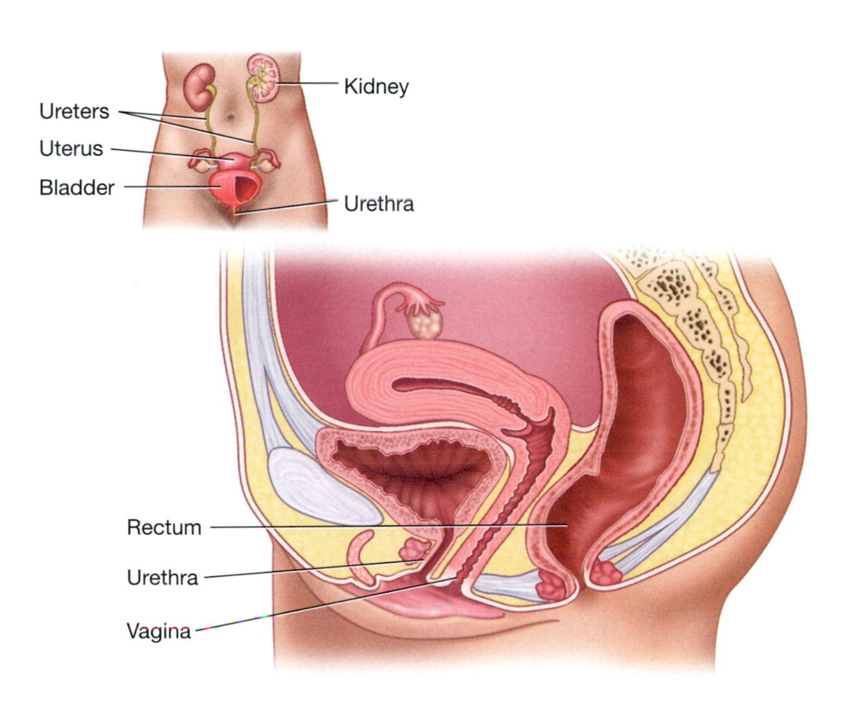

4 *How can this new problem be treated?*

A urine sample will be cultured in a Petri dish. Various antibiotic discs will be applied to the culture to determine the organism's antibiotic sensitivity. Antibiotics can cure most UTIs, but not all antibiotics that can kill *E. coli* in a Petri dish can eradicate a UTI. This is because not all antibiotics can concentrate at high enough levels

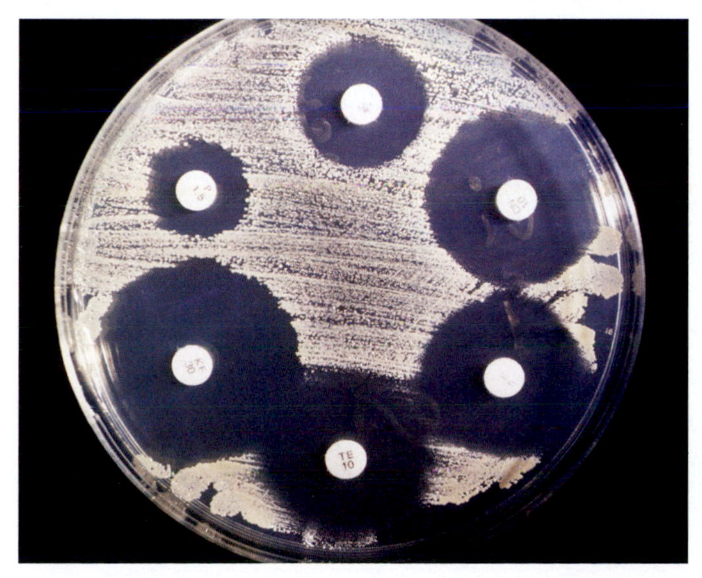

Credit: Biophoto Associates / Photo Researchers, Inc
Description: A bacterial antibiotic sensitivity test. Thin wafers containing different antibiotics are placed on an agar plate growing a bacterial culture. Bacteria are not able to grow around antibiotics when they are sensitive.

Link:
http://db2.photoresearchers.com/search/3Q6046
Contact info:
Photo Researchers, Inc.
60 East 56th Street, 6th Floor
New York, NY 10022
Tel: 212-758-3420
Fax: 212-355-0731
http://www.photoresearchers.com
info@photoresearchers.com

in the urinary tract to kill the bacteria. Examples of antibiotics that can concentrate in the urinary tract and are effective in treating most urinary tract infections are nitrofurantoin (Macrodantin) and the sulfa antibiotic combination of trimethoprim and sulfamethoxazole (Bactrim, Septra).

5 *Olivia's new problem resolves with treatment. But 3 weeks later her symptoms return. What is the likely cause of the return of her symptoms?*

Either Olivia has a new infection or she has a resurgence of the previous infection that might have developed resistance to the antibiotic used.

6 *Olivia is treated again and responds rapidly. Over the course of the next 6 months, her symptoms return three more times. Each time the same antibiotic leads to prompt resolution of her symptoms. What do you think is going on?*

For some reason urinary tract infections keep recurring in Olivia. They respond rapidly to the same antibiotic. The lab cultures indicate that the organism is *E. coli* each time and the bacteria are not resistant to the antibiotic.

A nursing student asks Olivia how she cleans herself after urinating. "Well young man—*that's* a very personal question! But I'll answer you. You know, having a shower only once a week in this place makes me feel dirty half the time. The Goodwin girls gave me a tip: After I go, I wipe and flush. Then I dip some toilet paper in the clean toilet water and wash myself."

The toilet water was *not* clean after flushing, and she was wiping her genital area with water contaminated with *E. coli.*

7 *Olivia has now been healthy for the last 4 months. However, her last urine culture was positive for bacteria—but she has no symptoms whatsoever. What treatment does she need?*

After Olivia changed her toileting habits, she has had no recurrence of UTI symptoms. Olivia now has asymptomatic bacteriuria, which means she has bacteria in her urine. Without symptoms, the current guidelines state that she should not be treated with an antibiotic. Evidence shows that in women with asymptomatic bacteriuria, the *E. coli* lose their virulence and actually protect against infection from more virulent strains.

References

Asymptomatic bacteriuria. In Chapter 100: Urinary Tract Infections, *Merck Manual of Geriatrics*:
http://www.merck.com/mkgr/mmg/sec12/ch100/ch100a.jsp

Todar, K. (2008). The Control of Microbial Growth. In: *Todar's Online Textbook of Bacteriology*. Madison: University of Wisconsin-Madison, Dept. of Bacteriology.
http://www.textbookofbacteriology.net/control.html

Urinary tract infection:
http://www.mayoclinic.com/health/urinary-tract-infection/DS00286

23

Suzie Feeds the Bunny

Laura invited her preschool class to her birthday party. Her pet Bunchy Bunny was put in his cage. Her 7-year-old brother, Jeremy, put a sign on the cage: "DO NOT FEED THE BUNNY." "That oughta do it!" He exclaimed.

"Laura," advised her mother, "Make sure that none of the kids put their fingers in the cage or Bunchy might bite them, thinking they're feeding him!"

When the children arrived, the first thing Laura did was to show them Bunchy.

"What does the sign say?" Suzie asked.

"Do not feed the bunny! And don't put your fingers in the cage or he will think you are food!" said Laura.

At that, Suzie immediately put her finger in the cage—and Bunchy bit her. "That hurt!" she cried, but then she ran off to see the clown who was making balloon figures.

Later that day Suzie's finger was red and swollen.

"What happened to your finger, Suzie?" her mother asked.

"I got a boo boo at Laura's party."

"How did you get the boo boo?"

"I dunno."

Suzie's mother put some triple antibiotic ointment on the wound, and topped it with a bright pink bandage.

"It feels okay, Mommy," Suzie assured her.

Four weeks later Suzie's finger seemed to be healed, but her finger was still warm to the touch with local edema (swelling) and erythema (redness). She complained that she had trouble holding a crayon. "I can't move this finger like my other fingers," she said. Suzie's mother became worried and brought her to the orthopedic clinic.

An X-ray of Suzie's finger was taken, after which she immediately was given an MRI of her legs and arms. An MRI (magnetic resonance image) is a diagnostic imaging technique whereby the patient is placed in a strong, uniform magnetic field. Protons absorb energy from the magnetic field and then emit radio waves as their excitation decays. These radio frequency signals are converted into three-dimensional images. MRIs do not expose patients to ionizing radiation.

Below is an X-ray of Suzie's finger:

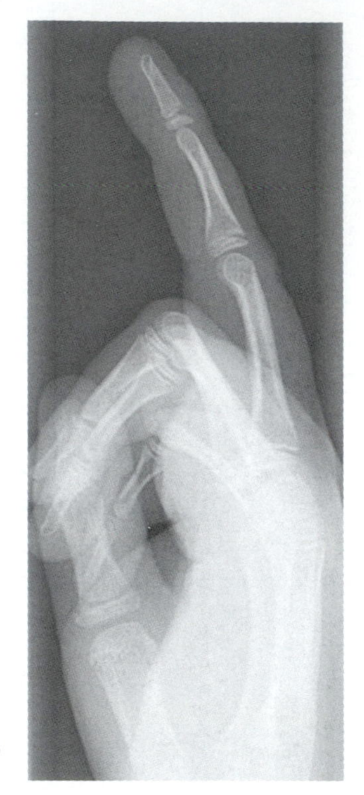

Courtesy of Sarah J. Fitch, Virginia Commonwealth University, Department of Pathology.

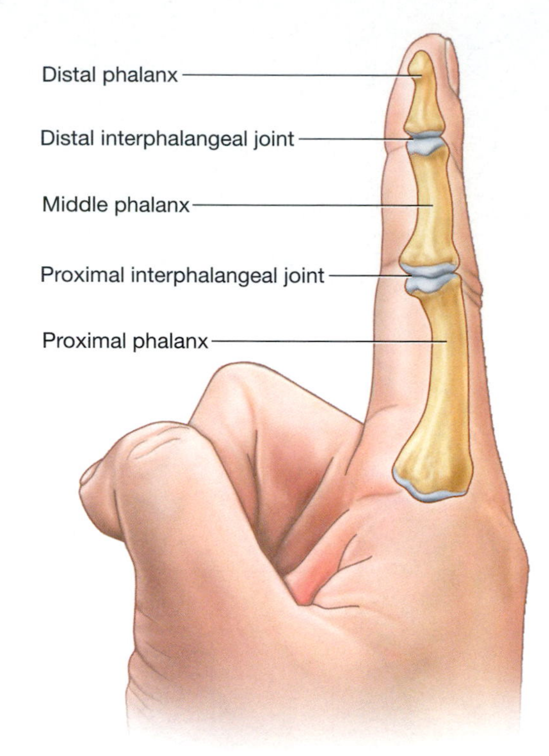

Questions

Name ______________________ Section ________

1 Find the dorsal aspect of the metaphysis of the distal phalanx on the X-ray. What do you see?

2 What condition does Suzie have?

3 Where else can this condition occur in the body?

4 How could this problem have been prevented?

5 What could happen if this problem is not treated?

6 Why was Suzie given an MRI of her legs and arms?

7 What else can cause this problem?

8 How can this problem be treated?

Answers to Questions

1 *Find the dorsal aspect of the metaphysis of the distal phalanx on the X-ray. What do you see?*

There is a region of osteolysis (destruction of the bone). At least 40% of focal bone loss must have occurred before lucency (transparency in the bone) can be seen in an X-ray.

2 *What condition does Suzie have?*

Suzie has "direct inoculation" or "contiguous-focus" osteomylelitis of the terminal phalanx. This is a bacterial infection of the bone secondary to inoculation of bacteria caused by direct contact during trauma. Bacteria from the rabbit's mouth gained entry to the bone, and multiple species of bacteria from the rabbit's mouth probably are involved. Typically, osteomyelitis does not show up on an X-ray for at least 2–4 weeks after it has occurred.

3 *Where else can this condition occur in the body?*

A penetrating wound in any part of the body near bone can cause osteomyelitis. The most common regions are shown in the figure below. Osteomyelitis is most common in children. Wounds to the femur and tibia are the most common sites of osteomyelitis. In children, however, the fingertips and toes also are common sites of osteomyelitis. Merely stubbing the finger or toe can cause the nail to penetrate the bone near the physis. This acts like an open fracture and allows bacteria to gain entry to the blood supply to the bone. Bacteria gaining entry to the blood from a seemingly trivial wound can migrate to the most vascular part of the long, growing bones such as the epiphysis in the femur.

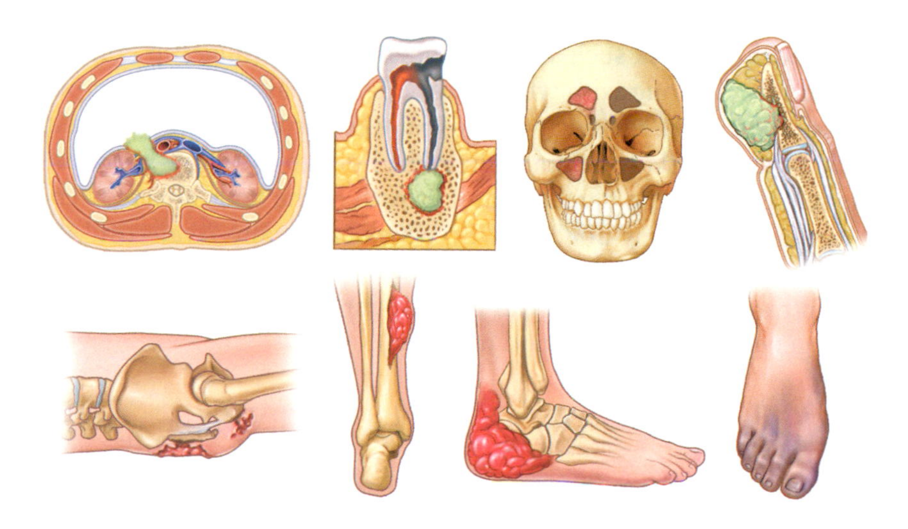

24

The Casino Bus

Mr. Jones, age 77, is a retired middle school history teacher. He has just been admitted as the first patient in the new hospital wing, paid for, in part, by two nearby casinos. Mr. Jones took the casino bus from his home, 3 hours away. He took a Benadryl so he could "sleep on the bus." After "playing the slots" for only 20 minutes, he began to have difficulty breathing. Someone called 911, and Mr. Jones was brought to the emergency department.

He reports that he has had "heart failure" and "COPD" but did not bring any medications or a list of his medications with him. His dyspnea responded to oxygen, an inhaled β_2 agonist (albuterol), and inhaled Atrovent (ipratropium bromide), a short-acting anticholinergic drug. His arterial blood gas values (ABGs) on admission are as follows:

Test	Value	Normal Range
PaO_2	55 mm Hg	75–100 mm Hg
SaO_2	88%	94%–98%
$PaCO_2$	46 mm Hg	35–45 mm Hg
pH	7.3	7.35–7.45
HCO_3	30	22–26 mEq/L

He has +4 pedal edema, sometimes called "pitting edema." Shown here is a photograph of his foot.

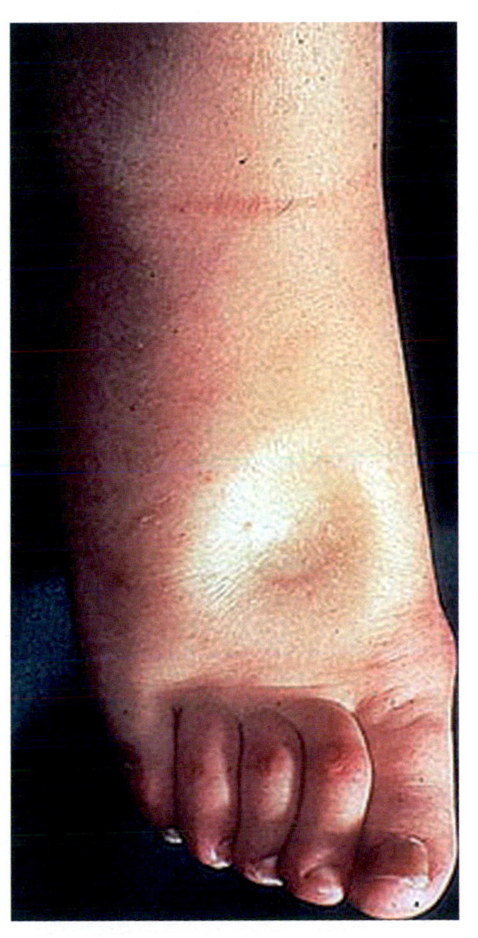

Reprinted with permission of the Department of Pathology, Virginia Commonwealth University and the VCU Health System

His ECG, shown below indicates he has left ventricular hypertrophy. An echocardiogram confirmed that he has concentric left ventricular hypertrophy and a ventricular ejection value < 60%.

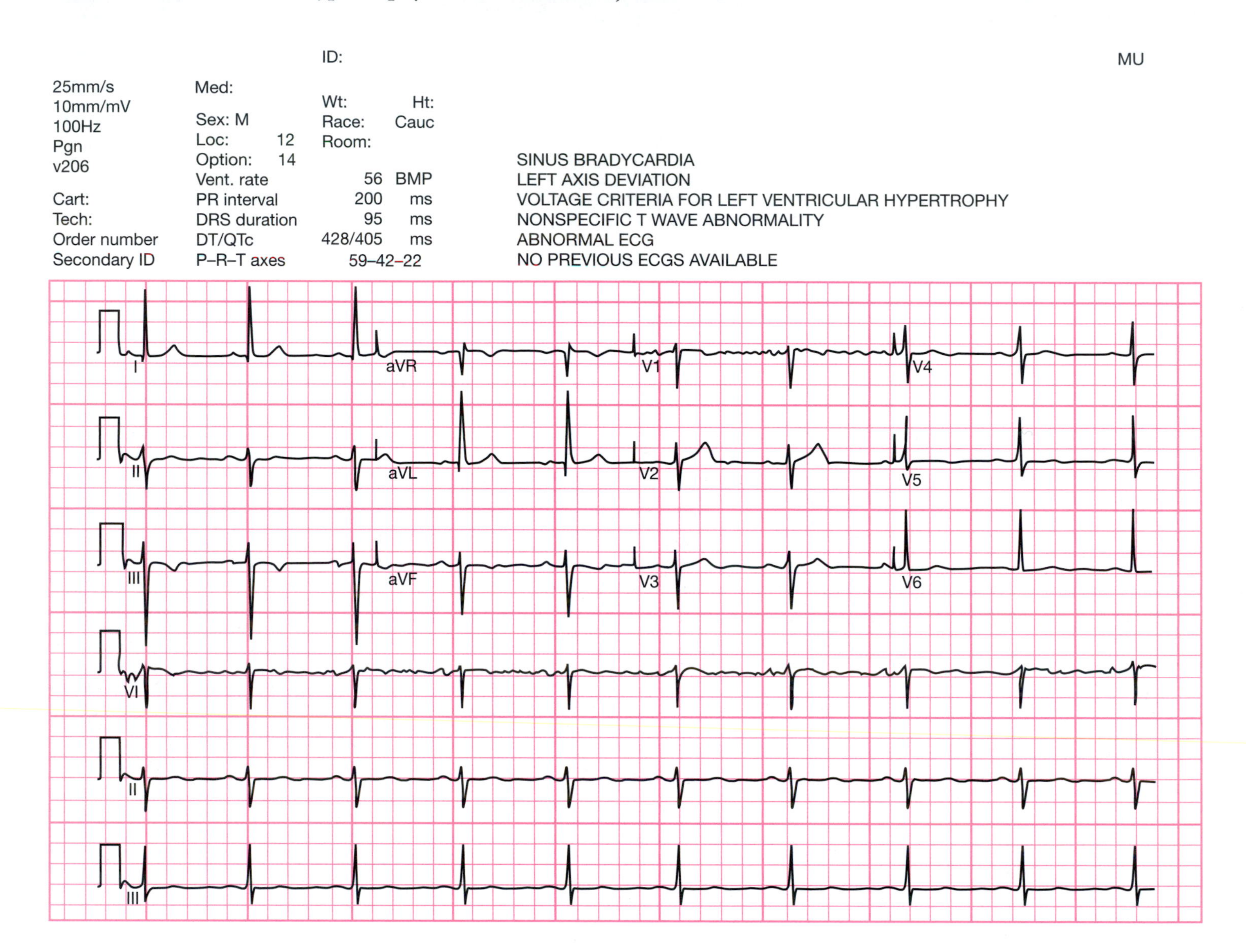

A normal ECG is shown below:

ID: MU

25mm/s	Med:	Wt: 156lb	Ht:
10mm/mV		Race:	Cauc
100Hz	Sex: M	Room:	
Pgn	Loc: 12		
v206	Option: 14		
	Vent. rate	73	BMP
Cart:	PR interval	137	ms
Tech:	QRS duration	80	ms
Order number:	QT/QT=	388/426	ms
Secondary ID:	P–R–T axes	28–34–58	58

NORMAL SINUS RHYTHM
NORMAL ECG

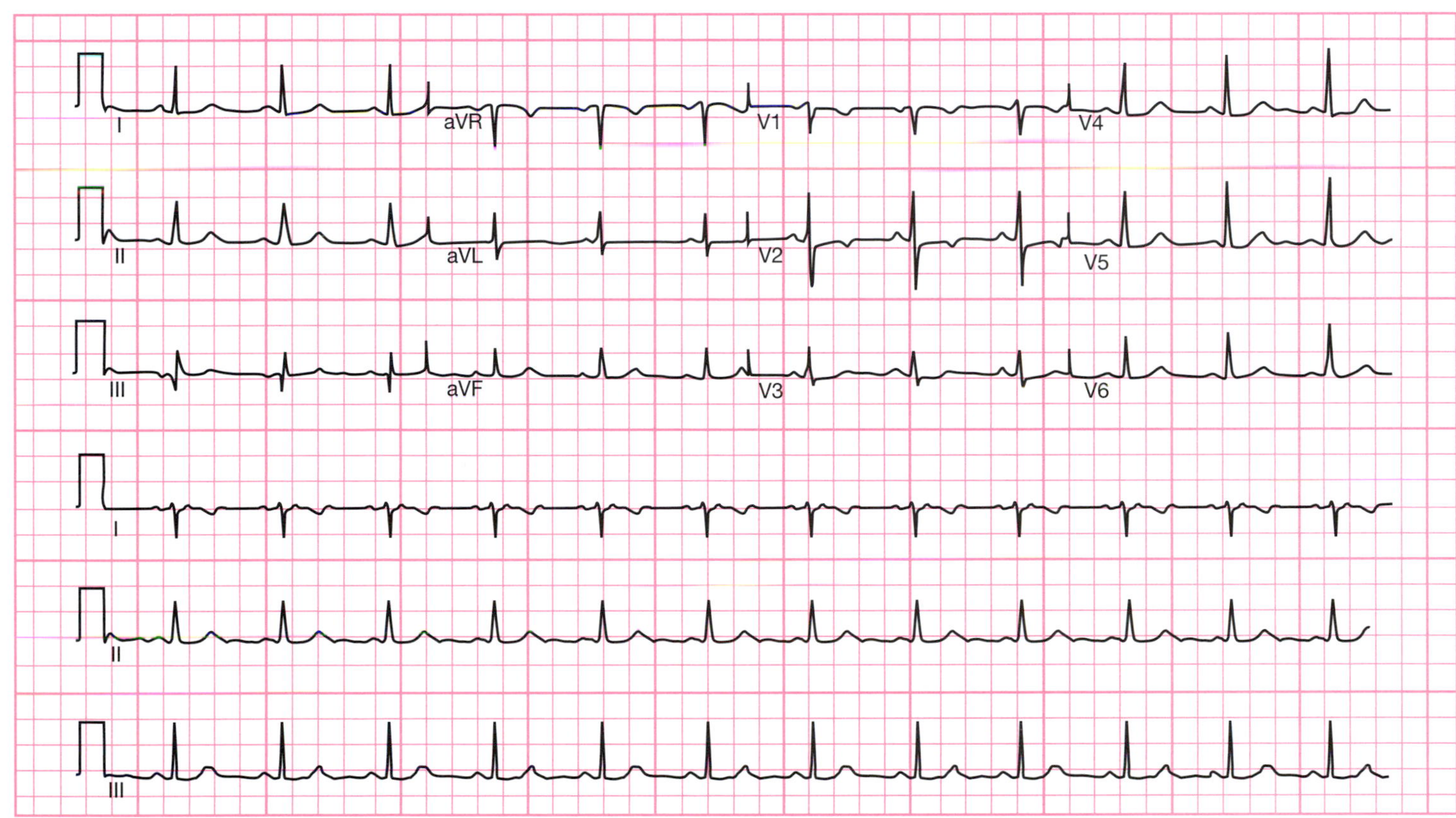

Mr. Jones is given furosemide (Lasix), a diuretic medication that reduces water and sodium reabsorption from the loop of Henle. It is used for patients with left-sided heart failure.

When asked how he feels right now, he says, "Thanks—I feel okay, but I can't go. I tried this urinal thing, but nothing is coming out. And now I have a pain in my lower stomach."

The nurse catheterizes Mr. Jones with a Foley catheter and urine begins to flow into the catheter bag. "Usually I have to go all the time," says Mr. Jones, "but then when I do go—it starts and then it stops and I just cannot empty my bladder! I hate this tube thing, but it sure relieves the pain in my bladder!" Later that day Mr. Jones is given a digital rectal exam and a PSA (prostate-specific antigen) blood test. He learns that he has benign prostatic hypertrophy (BPH).

Questions

Name ______________________________ Section __________

1 What is COPD? What are the symptoms? What causes it?

2 How are chronic bronchitis and emphysema related to COPD?

3 An agonist drug binds to the drug receptor and mimics the actions of an endogenous substance (naturally occurring in the body). Where are the $ß_2$ receptors in the body and what is the endogenous substance that binds to them? What happens as a result?

4 Anticholinergic drugs bind to cholinergic receptors. Acetylcholine is the endogenous substance that binds to cholinergic receptors. But inflammatory mediators (histamine, bradykinin, eicosanoids) also stimulate cholinergic receptors. What happens when cholinergic receptors are stimulated in the airways? Why would oral anticholinergic drugs such as the over-the-counter antihistamine diphenhydramine (Benadryl) be dangerous for a patient with COPD but the inhaled anticholinergic drug Atrovent (ipratropium) be beneficial?

5 What do the arterial blood gas values suggest?

6 What causes left-sided heart failure? What symptoms of heart failure does Mr. Jones have?

7 Lasix is a diuretic medication that reduces water reabsorption from the loop of Henle. It is used with patients who have left-sided heart failure. Why would a diuretic affecting the loop of Henle help Mr. Jones?

8 How does a Foley catheter "work?"

9 What are the symptoms of BPH, and what causes it?

10 What did Mr. Jones do to cause his present symptoms?

Answers to Questions

1 *What is COPD? What are the symptoms? What causes it?*

COPD stands for chronic obstructive pulmonary disease. The hallmark of the disease is a progressive reduction in airflow in the lungs. Patients with COPD respond with an exaggerated inflammatory response in the lung when exposed to noxious gases or airborne particles (e.g., dust, cigarette smoke, ozone, and other air pollutants). Symptoms are chronic cough with production of sputum—often for years. Eventually patients develop reduced airflow and suffer dyspnea on exertion. Most patients have episodes of worsening symptoms.

Risk factors are tobacco smoke (smoking or environmental tobacco smoke), intense or prolonged exposure to dusts or irritating vapors/fumes, indoor and outdoor air pollution. Low birth weight and childhood respiratory infections increase the risk that a smoker will develop COPD.

2 *How are chronic bronchitis and emphysema related to COPD?*

A patient is said to have chronic bronchitis when he/she has a cough with sputum production (i.e., "productive cough") for at least 3 months in each of 2 consecutive years. But these patients do not necessarily have reduced airflow, so may not be classified yet as having COPD. Their risk of developing COPD eventually is high, however.

Emphysema involves the destruction of the alveoli. Emphysema is just one of several possible lung abnormalities in patients with COPD. Patients with emphysema have COPD, but not all patients with COPD have emphysema.

3 *An agonist drug binds to the drug receptor and mimics the actions of an endogenous substance (naturally occurring in the body). Where are the β_2 receptors in the body, and what is the endogenous substance that binds to them? What happens as a result?*

The β_2 receptors are in the lung, and the heart has β_1 receptors. (The way to remember this is that "there are 2 lungs and 1 heart".) Stimulation of the β_2 receptors in the lung causes bronchodilation. The endogenous substance that binds to them is epinephrine. Epinephrine is not selective for β_2 receptors, though. It also binds to β_1 receptors in the heart, causing an increased heart rate, and it binds to $alpha_1$ receptors on arterial endothelium, causing vasoconstriction.

Using epinephrine is fine for severe allergic reactions (anaphylaxis) because it will cause bronchodilation and elevate blood pressure. But if used in dyspnea with COPD, it would aggravate underlying cardiac disease. Instead, *selective* β_2 agonists such as inhaled albuterol are used to stimulate bronchodilation and alleviate dyspnea in COPD.

4 *Anticholinergic drugs bind to cholinergic receptors. Acetylcholine is the endogenous substance that binds to cholinergic receptors. But inflammatory mediators (histamine, bradykinin, eicosanoids) also stimulate cholinergic receptors. What happens when cholinergic receptors are stimulated in the airways? Why would oral anticholinergic drugs such as the over-the-counter antihistamine diphenhydramine (Benadryl) be dangerous for a patient with COPD but the inhaled anticholinergic drug Atrovent (ipratropium) be beneficial?*

The nervous system is composed of the central nervous system (CNS) and the peripheral nervous system (PNS). The PNS is divided as follows:

Peripheral nervous system

- Somatic motor system
- Autonomic nervous system
  - Parasympathetic nervous system
    - contracts bronchial smooth muscle
    - stimulates secretion of mucus in airways
    - slows heart
    - increases gastric acid secretion
    - controls bladder and bowel
    - focuses eye and constricts the pupil
  - **Sympathetic nervous system**
    - opposes effects of parasympathetic nervous system
    - controls cardiovascular function (heart rate, vasoconstriction)
    - regulates body temperature (dilation, constriction of cutaneous blood vessels; promotes sweating, piloerection)
    - initiates sympathetic stress response ("flight or fright")

In most organs the effects of the parasympathetic nervous system oppose the effects of the sympathetic nervous system. Cholinergic receptors are important in both! There are two subtypes of cholinergic receptors: muscarinic and nicotinic. Muscarinic receptors are on the sweat glands (sympathetic nervous system) and in all the organs innervated by the parasympathetic nervous system. Nicotinic receptors are on the postganlionic neurons of both the parasympathetic nervous system and the sympathetic nervous system, as well as the adrenal medulla.

Stimulation of muscarinic cholinergic receptors in the airways stimulates secretion of mucus. Inhaled anticholinergic drugs (such as Atrovent) block muscarinic cholinergic receptors in the airways and reduce the production of mucus.

Think of the three-letter word "dry" when you consider the systemic effects of an anticholinergic drug that binds to nicotinic cholinergic receptors in the sympathetic nervous system. Benadryl (diphenhydramine) is an antihistamine (blocks histamine receptors), but it also has *systemic* anticholinergic effects. It can cause sedation, reduce urination (anticholinergic effects on bladder), and can dry up respiratory secretions. The latter is dangerous in patients with COPD because it can cause the mucus to form dry "plugs" that can restrict the airways.

Inhaled anticholinergic drugs such as Atrovent (ipratropium) are now first-line drugs used to treat COPD. These are very special molecules that selectively block the ability of certain inflammatory mediators (histamine, bradykinin, eicosanoids, cigarette smoke, sulfur dioxide, ozone) to stimulate muscarinic cholinergic receptors in the central airway wall and the submucosal glands. These molecules don't dry up the mucus; they prevent excess mucus from being produced in the first place. They don't cross the blood-brain barrier and have minimal systemic effects.

5 *What do the arterial blood gas values suggest?*

Test	Value	Acid Range	Normal Range	Alkaline Range
PaO_2	55 mm Hg		75–100 mm Hg	
SaO_2	88%		94%–98%	
$PaCO_2$	46 mm Hg	> 45	35–45 mm Hg	< 35
pH	7.3	< 7.35	7.35–7.45	> 7.45
HCO_3	30	< 22	22–26 mEq/L	> 26

Upon arriving at the ED, Mr. Jones had respiratory acidosis. When airflow into the lungs is impaired in COPD, the partial pressure of oxygen (PaO_2) level in the blood decreases. Less oxygen is available to bind to hemoglobin and be transported to tissues. Mr. Jones's percent of hemoglobin saturated with oxygen ("oxygen saturation," or SaO_2) was low upon admission. Because he was not meeting his oxygen needs, he was given oxygen by nasal canula. His CO_2 level in the blood was elevated slightly.

In COPD, CO_2 transport across alveoli membranes is much more efficient than oxygen transport, so the degree of carbon dioxide elevation usually is less than oxygen reduction. The excess blood CO_2 combines with water in the blood to form carbonic acid. This lowers the pH of the blood, so an elevation in $PaCO_2$ is considered a sign of acidosis.

Bicarbonate ion (HCO_3) is an alkaline molecule that can bind to free H^+ and bring the pH back up to normal. This is a slow process. In the meantime, the excess carbonic acid causes an increased respiratory rate in a physiological attempt to exhale the excess CO_2. The patient with COPD experiences dyspnea (difficulty breathing) because of the impaired airflow. The kidneys can compensate also by increasing the excretion of hydrogen ions and increasing the reabsorption of bicarbonate. (But Mr. Jones had not urinated.)

6 *What causes left-sided heart failure? What symptoms of heart failure does Mr. Jones have?*

The most common cause of left-sided heart failure is poorly controlled hypertension. Hypertension increases the peripheral resistance of the arteries, causing arterial constriction. This increases the "afterload," which is the amount of resistance to blood ejected by the left ventricle. The heart then has to work harder and eventually fails. Other causes of left-sided heart failure include coronary atherosclerosis, valve defects, and heart attack (clot in a blood vessel that supplies oxygen to the heart muscle). Failure of the left heart to pump efficiently reduces blood flow to the kidneys and blood backs up in the lungs. As a result, patients experience reduced urine output, edema in the extremities, and pulmonary congestion, causing difficulty breathing (dyspnea), all of which Mr. Jones is exhibiting.

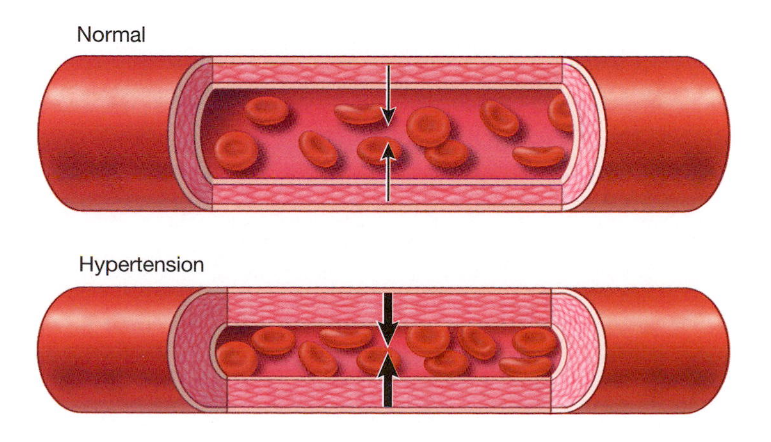

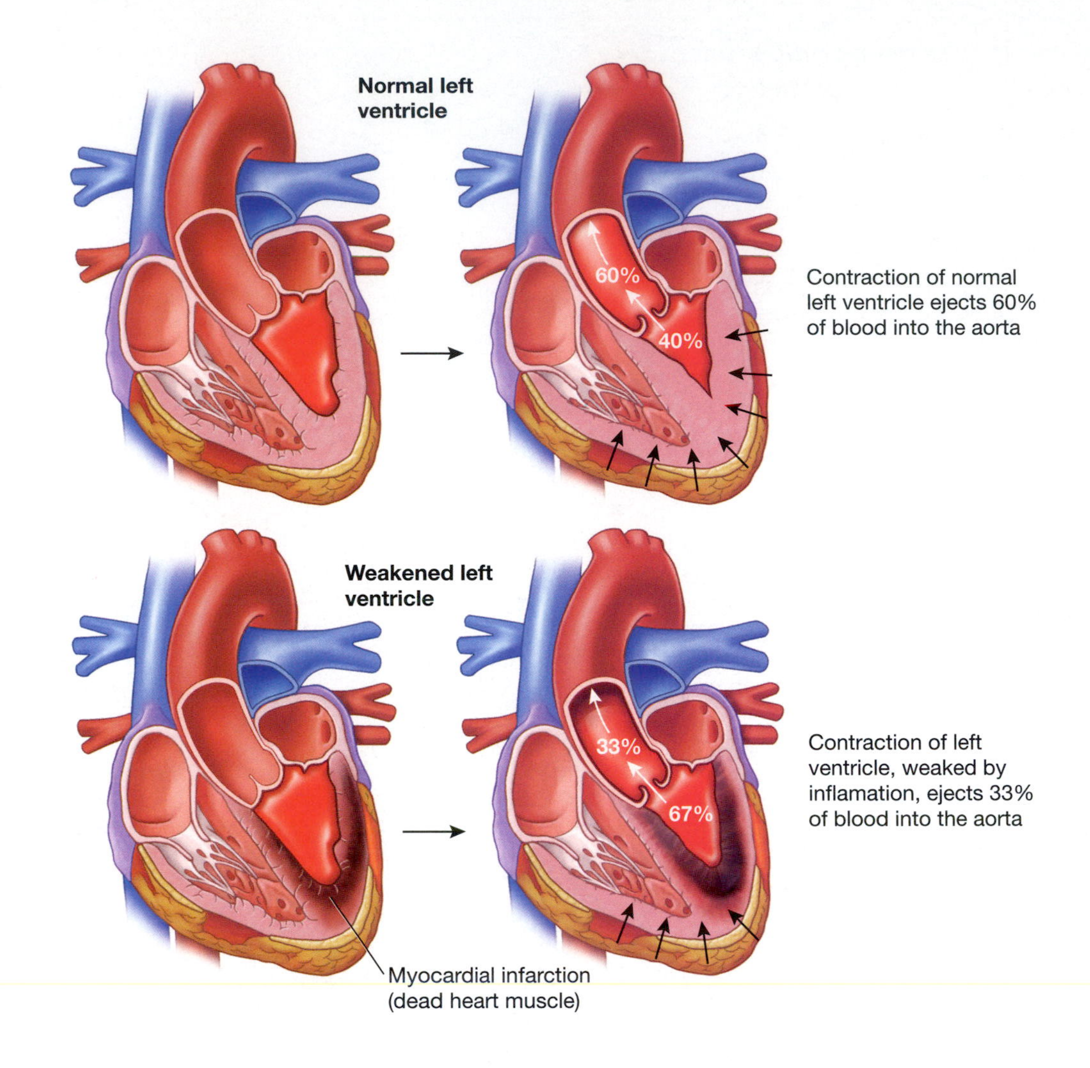
Normal left ventricle
60%
40%
Contraction of normal left ventricle ejects 60% of blood into the aorta
Weakened left ventricle
33%
67%
Contraction of left ventricle, weaked by inflamation, ejects 33% of blood into the aorta
Myocardial infarction (dead heart muscle)

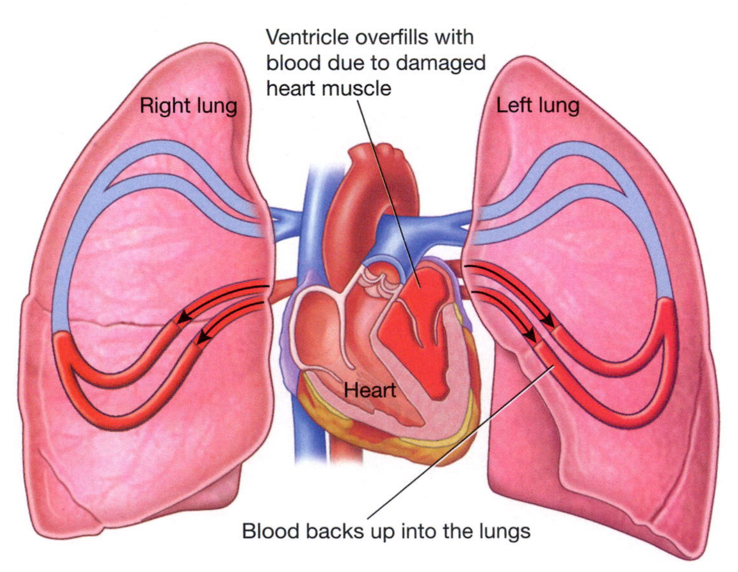
Ventricle overfills with blood due to damaged heart muscle
Right lung
Left lung
Heart
Blood backs up into the lungs

7 *Lasix is a diuretic medication that reduces water reabsorption from the loop of Henle. It is used with patients who have left-sided heart failure. Why would a diuretic affecting the loop of Henle help Mr. Jones?*

Lasix (generic name = furosemide) works by inhibiting the chloride pump in the ascending loop of Henle, which causes active transport of chloride (and sodium along with it). As a result, more chloride and sodium move along the tubule and end up in urine. The extra chloride and sodium in the filtrate inhibits osmotic passive diffusion of water out of the descending loop, so more water is eliminated. Obviously, diuretics should never be administered to dehydrated patients.

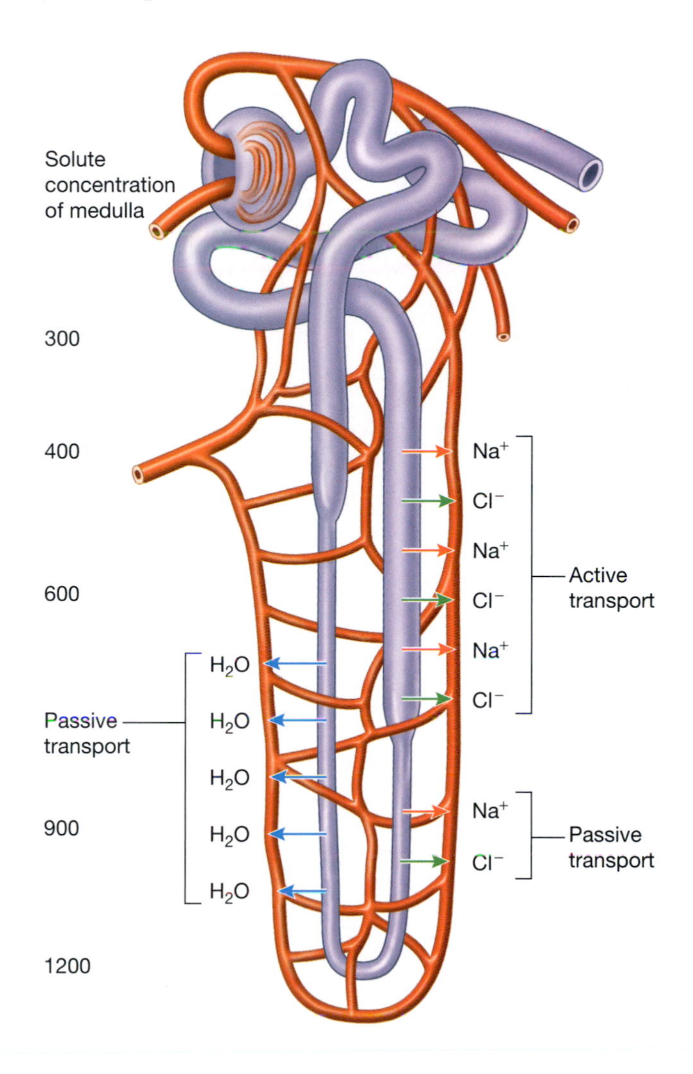

Countercurrent Mechanism

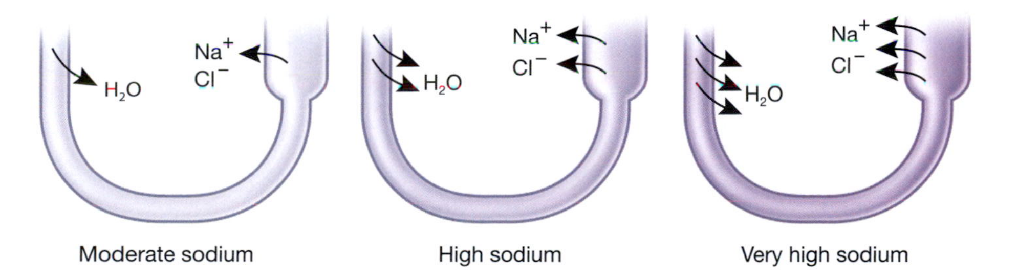

8 *How does a Foley catheter "work?"*

A Foley catheter is a hollow tube with an inflatable balloon at the tip. The catheter is inserted through the urethra and into the bladder. Sterile water is inserted into the catheter and the balloon expands. This prevents the catheter from falling out. Urine drains through the upper hole in the catheter.

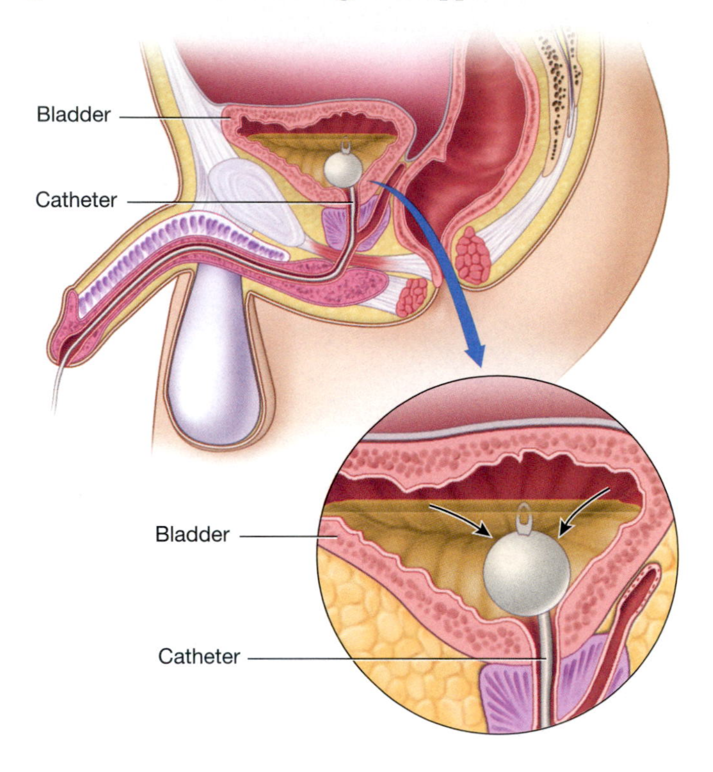

9 *What are the symptoms of BPH, and what causes it?*

Benign prostatic hypertrophy (BPH) is the enlargement of the prostate gland, which surrounds the urethra in the male. Half of men aged 50 and older have BPH. By age 80, 90% of men have BPH. It does *not* increase

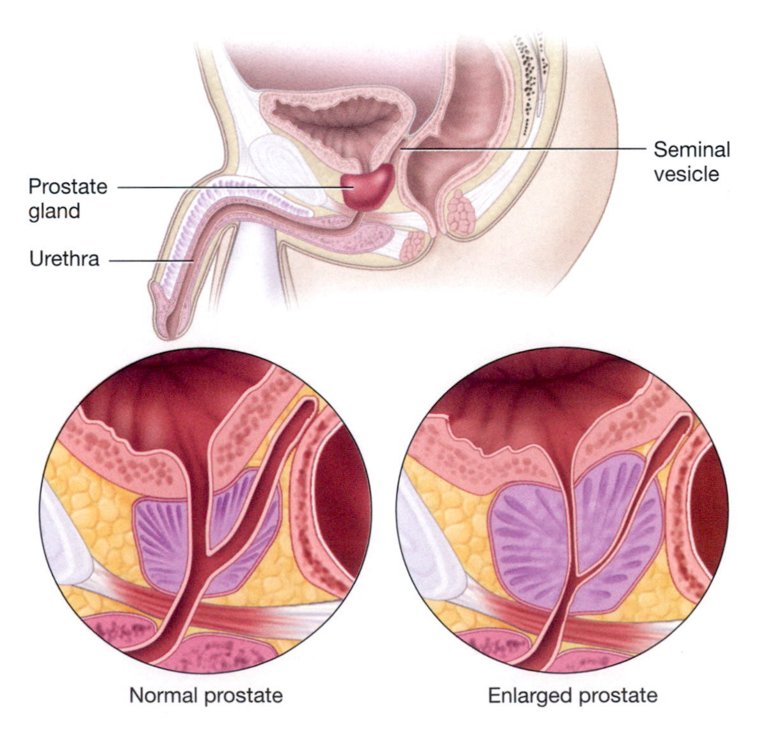

the risk of prostate cancer, but cancer should be ruled out with the diagnosis and PSA levels monitored thereafter. An elevation in PSA may signal the development of prostate cancer.

As the prostate grows, the prostate constricts the urethra. Symptoms of BPH include a frequent urge to urinate, getting up to urinate more than once a night, urination that starts and stops, and difficulty emptying the bladder. Mr. Jones has acute urinary retention, which was relieved with the catheter. He will have to start on a medication that shrinks the prostate over time. His symptoms will improve within about 3 months.

10 *What did Mr. Jones do to cause his present symptoms?*

Mr. Jones said, "You know, I should have told them this downstairs, but I didn't take my Lasix or my other heart pills yesterday or this morning. It's too hard to line up on the bus to use the toilet. I'm too unsteady on my feet. So a lot of us have learned that if we just skip the Lasix and don't drink anything, we can get to the casino without having to go. I take the Benadryl so I can sleep on the bus. When I got to the casino, I was so excited because I won on the first machine. I thought my excitement was what made me short of breath!"

Skipping the Lasix and other medications for his heart failure resulted in a failure of Mr. Jones' left ventricle to eject sufficient blood to perfuse the kidneys, so less urine was produced. This caused the pitting edema in his feet and contributed to his dyspnea. The systemic anticholinergic effects of Benadryl aggravates BPH (inhibits bladder contraction) and can cause mucus plugs in patients with COPD.

Following insertion of the catheter and administration of fluids followed by Lasix (and his other heart medications), Mr. Jones was soon producing urine on his own—not the forceful stream he would have in 3 months after his BPH medicine takes effect, but enough to be discharged without a catheter. Mr. Jones is given a medication record that folds up neatly and fits in his wallet.

"I'll be back in the spring," he says, but next time I'll take my meds and take this record everywhere I go! I guess I have some teaching to do with my friends!"

References

BPH Guidelines
http://www.auanet.org/content/guidelines-and-quality-care/clinical-guidelines.cfm?sub=bph

Global Initiative for Chronic Obstructive Lung Disease
http://www.goldcopd.com/

Heart Failure Site for Patients
http://www.heartfailure.org/
http://www.americanheart.org/presenter.jhtml?identifier=1486

Heart Failure Treatment Guidelines
http://www.heartfailureguideline.org/

Index